Epidemic!

ALSO IN THE SERIES

The Biodiversity Crisis: Losing What Counts
edited by Michael Novacek

Epidemic!

The World of Infectious Disease

DR. ROB DESALLE, EDITOR

published in conjunction with

The American Museum of Natural History

THE NEW PRESS

Computer-generated image of Hepatitis B virus

Published in the United States by THE NEW PRESS, NEW YORK
Distributed by W. W. NORTON & COMPANY, INC., NEW YORK

The New Press was established in 1990 as a not-for-profit alternative to the large, commercial publishing houses currently dominating the book publishing industry. The New Press operates in the public interest rather than for private gain, and is committed to publishing, in innovative ways, works of educational, cultural, and community value that are often deemed insufficiently profitable.

www.thenewpress.com

Printed in the United States of America

9 8 7 6 5 4 3 2 1

Contents

SECTION 3. Infection

SECTION 4. Outbreaks

19.95

SECTION 5. Epidemics and Pandemics

SECTION 6. Action

Foreword ELLEN V. FUTTER

Since its founding in 1869, the American Museum of Natural History has put the world on display. The twin pillars of our mission have always been science and education. Today, the Museum is one of the world's leading research institutions in the natural sciences. Over the years our scientists, now a staff of more than 200 men and women, have gone on more than 100 expeditions a year, collecting evidence from all over the globe in their effort to answer questions about such fundamental scientific and human issues as the origins of the universe, earth, and life, and who we are and where we fit. In addition to research, our scientists have a related responsibility—to interpret science for the general public. The exhibitions at the Museum have been conceived and curated by scientists who are committed to putting the evidence—"the real stuff"—in front of the public.

As we move into the twenty-first century, we at the Museum are filled with a renewed dedication to our mission. To many people, science today seems too removed and too difficult to understand, yet the need has never been greater for a public that is well informed about science. Through its educational initiatives and exhibitions, the Museum seeks to narrow the gap between what people know and what they need to know about science. To that end, in 1997 the Museum launched the National Center for Science Literacy, Education, and Technology to extend the Museum's resources beyond its walls to a national audience.

Science is exploration. Scientists work at the frontier—at the border of the known and the unknown. This new book series, through the words of working scientists, enables non-scientists to share the excitement of cutting-edge science. The series will begin with four volumes that expand the themes covered in many of our major new exhibitions. Our exhibitions always embody a scientist's vision and point of view. In the same way, each book in this series is "curated"—researched, organized, and introduced—by one of the Museum's scientists. Each book features a selection of essays written by leading scientists who have made significant contributions to the field. The essays are supported by case studies and profiles of important people and events.

The series begins with *Epidemic! The World of Infectious Disease*, a book created to explore the themes of "Epidemic! The World of Infectious Disease", the major new traveling exhibition. This exhibition looks at infectious disease from a natural history perspective, and at the complex relationship between ecology, evolution, and culture.

The series continues with three books to support our new permanent halls at the Museum. In May 1998, the Museum opened the Hall of Biodiversity to great critical acclaim. *The Biodiversity Crisis: Losing What Counts* will address the main themes

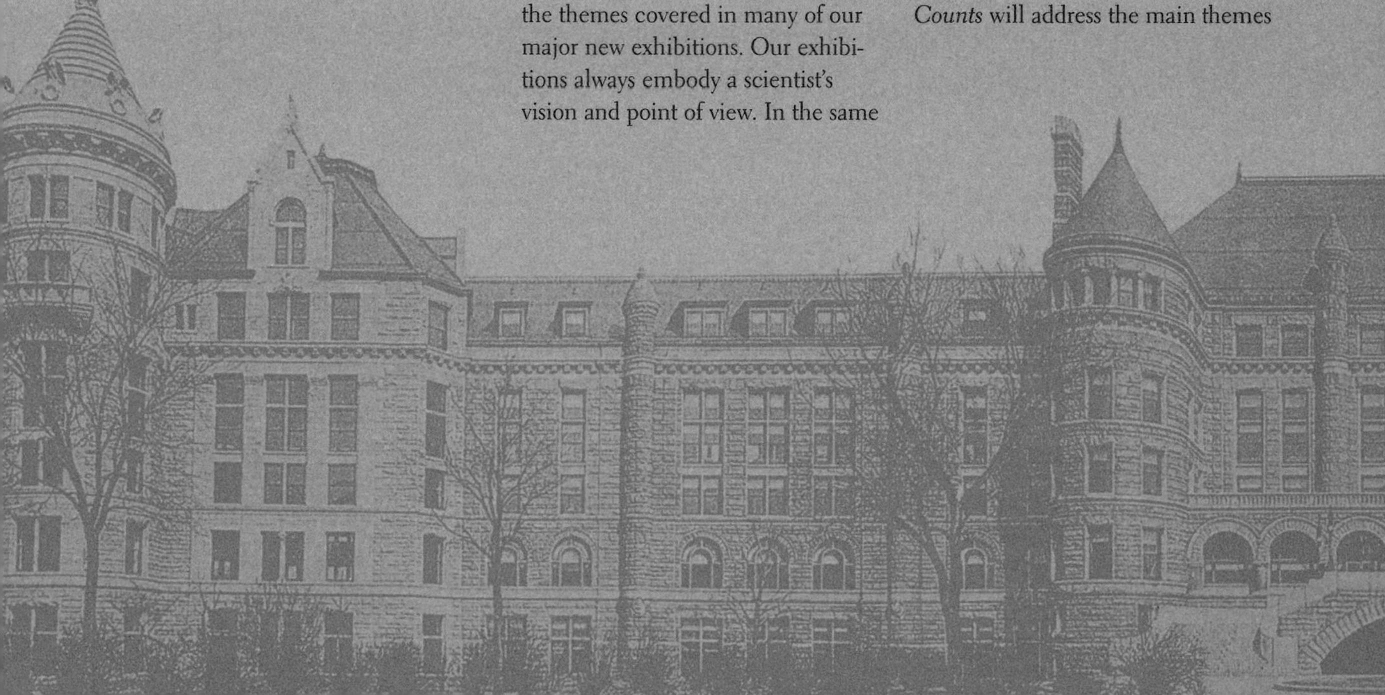

of that Hall, focusing on the variety and interrelatedness of all living things on our planet. Other books will focus on the Earth and the phenomena that shape our planet to expand the themes addressed in our new Hall of Planet Earth, and the mysteries and wonders of the universe to expand the themes addressed in our new Hall of the Universe.

This new series illustrates our continuing commitment to connect the general public with the natural world. We cannot send real specimens to every home and classroom, but we can bring the ideas, concerns, and questions of working scientists directly to you. We hope this series provides a valuable resource that will prepare tomorrow's leaders to make informed decisions about the world that we all share.

ELLEN V. FUTTER,
President, American Museum of Natural History

Acknowledgments

This book was made possible by a generous grant from an anonymous foundation. It was developed in conjunction with "Epidemic! The World of Infectious Disease", an exhibition at the American Museum of Natural History, made possible through the generous support of the Bristol-Myers Squibb Company. Major funding for the exhibition has also been provided by the Lila Wallace-Reader's Digest Endowment Fund.

This book was produced by the National Center for Science Literacy, Education, and Technology, American Museum of Natural History.

We would like to thank Ellen Reeves, Barbara Chuang, and Grace Farrell at the New Press, along with Hall Smyth and Steve Reinke at BAD.

EDITORIAL AND PRODUCTION STAFF:

Producer and Senior Editor:
CAROLINE NOBEL

Science Editor:
VARUNI KULASEKERA, Ph.D.

Education Editor:
NATHAN CHU

Writers:
THERON COLE, JR.,
BARBARA RAVAGE

Production Coordinators:
FRANCINE MILLMAN,
ELLEN PRZYBYLA

Production Assistant:
LETTICE NESBIT

ELLEN V. FUTTER, *President*

MYLES GORDON, *Vice President of Education*

NANCY HECHINGER,
Director of National Center for Science Literacy, Education, and Technology

photo: Exterior of the American Museum of Natural History—circa 1900.

Bust of Louis Pasteur that resided in the American Museum of Natural History's Hall of Public Health in the early twentieth century.

Introduction: Natural History of Infectious Disease

ROB DESALLE

"It's time that we close the book on infectious disease."

—WILLIAM STEWART, *Surgeon General*, United States, 1967

A great seventeenth-century invention, the microscope, revealed unimaginably strange "microscopic" creatures with which humans share the earth. In the late nineteenth century, brilliant scientific investigations established that some of these microbes were the cause of infectious disease. From then on, the possibility of conquering infectious disease, by controlling these microbes, has tantalized humanity.

The first attempts to control microbes were usually mechanical, such as the now-famous pasteurization process. They were based on an understanding of the lifestyles of the microbes causing the maladies. Moreover, they scored some striking successes; the pasteurization of milk, for example, greatly reduced the incidence of tuberculosis. In the 1870s, Pasteur, emboldened by these successes, suggested that the book on disease could be closed: "It is in the power of man to make parasitic maladies disappear from the face of the globe." Here is the earliest known assertion that humans could eradicate disease by exploiting knowledge about how microbes live and reproduce.

Pasteur's bold prediction resounded as late as 1967. The then U.S. Surgeon General, William Stewart, specifically chose the image of "closing the book on infectious disease" when he publicly predicted that the fruits of nearly thirty years of antibiotic development, together with the advances expected to flow from the new science of molecular biology, might finally lead to disease eradication. Yet, today, after another thirty years of some of the most fantastic developments in molecular biology—DNA cloning, the polymerase chain reaction, DNA sequencing, the human genome project, the determination of the entire genomes of microbes,

Rob DeSalle is the Co-Director of the Molecular Laboratories and an Associate Curator at the American Museum of Natural History where he curated "Epidemic! The World of Infectious Disease."

to name just a few—the book on infectious disease appears more wide open than in Pasteur's time, even than in Stewart's time.

In the 120 years since Pasteur's suggestion that medicine could wipe parasitic maladies from the face of the earth, hundreds of millions of people have died of infectious diseases—tuberculosis, AIDS, malaria, dengue, smallpox, cholera, plague, influenza, and hundreds of other diseases. Why has Pasteur's hope—and the hopes of so many scientists following him—been unfulfilled? The answer appears to lie in the failure to understand the ecological and evolutionary interactions we humans have, and have always had, with microbes. So intimate and essential is our relationship with microbes that the book on infectious disease will remain open for as long as we exist as a species. Our goal then must be to read the book more closely—to study it so that we can more carefully understand the intricate balancing act of man and microbe, health and disease. In this process, we will learn much more about the natural history of infectious disease—about why it exists, and about how we might control it.

To envision the coexistence of humans and microbes in a particular ecosystem, picture a seesaw in the playground. The seesaw is the ecosystem. We sit on one side, with our particular biological adaptations, and on the other side are the microbes, with their biological adaptations. In the playground, the seesaw will not balance unless you and your partner on the opposite seat are the same weight. If the two of you are roughly the same weight, balancing is easy. However, if your partner weighs fifty pounds more than you do, what do you do? You go through a series of adjustments or adaptations. You may vary your position on the seesaw, or move the fulcrum of the seesaw, or maybe place your book bag right in front of you to give your end more weight; perhaps you simply wait for someone closer to your weight to show up.

In a stable ecosystem, the seesaw remains essentially in balance. Normal fluctuations in the ecosystem may cause one end to swing up—causing the other end to drop—but ecosystems can return to balance. What happens, however, when an abnormal fluctuation such as a flood, a war, a failed vaccination program, or a mutation in a once-benign virus occurs? One end of the ecological seesaw swings up, quickly and sharply, and remains in the air, while the other end slams into the ground. Then, restoration of balance requires a more substantial series of adjustments by both humans and microbes—a more substantial repositioning by both on the ecological seesaw.

In the playground, knowledge of the physics of seesaws can tell you what will work and what will not work, when you attempt to regain a balance. Similarly, by knowing the life cycles and habitats of vectors—mosquitoes, fleas, ticks, etc.—that with a single bite can transfer a disease-causing parasite or virus into our bloodstream, we may find ways to decrease infectious disease. A little bit of knowledge might tell us that spraying vast areas with pesticides to kill all insect vectors is a good solution, but an understanding of these organisms would show that this is shortsighted. Evolution by the vectors will cause the pesticides to lose their potency. For example, DDT-resistant mosquitoes can appear after only two growing seasons.

Microbes adapt incredibly rapidly and efficiently because they multiply and evolve so quickly. Although humans and other organisms (plants and animals) can adapt to microbial attempts to "parasitize" them, these adaptations occur much more slowly than do those of microbes. When humans adapt, microbes then make new evolutionary adjustments that human populations must cope with. Evolutionary biologists have often called this leapfrogging process an arms race. Sometimes microbial-human interactions end in seemingly peaceful and mutually beneficial coexistence, as in our relationship with the *E. coli* bacteria that produce vitamins for us while luxuriating in our large intestine. Sometimes these interactions result in a hostile stalemate, as between the Ebola virus and us. Ebola is such an efficient killer that its human victims die too quickly to spread it from its limited range in Africa. Other times, the arms race between humans and microbes continues endlessly, as in the struggle between humans and the bacilli causing tuberculosis. Periods in which tuberculosis has been a major human scourge have alternated with ones of relative human immunity to it. After a recent period during which antibiotics controlled tuberculosis, now it is again resurgent. We may well bring tuberculosis under control again but its eventual total control depends on infectious disease scientists continuing to keep the lead in our arms race with it.

Another extremely important factor of the arms race between microbes and humans involves a characteristic unique to humans. Human responses to infectious disease include cultural responses, that is, political, economic, social, and behavioral responses. Human culture thus becomes an important aspect of the natural history of infectious disease—both affecting it and being affected by it. Because culture is so important to the story of infectious disease, this book includes several definitions of infectious disease; these definitions differ because they are based on differing cultural and social perspectives. The differences in definition and interpretations of the natural history of infectious disease could hamper or help the global fight against infectious disease—which is challenging us in both ecological and evolutionary contexts. How well we can reconcile these differences may determine our future success in the investigation, treatment, and prevention of infectious disease.

We must first direct our attention to local issues of natural history. We must understand ecology. This requires more than understanding the molecular biology and genetics of microbes, of vectors, and of ourselves; it also requires understanding the human interactions with vectors and with microbes. We must understand the environmental factors that tip the balance in these human-vector-microbe interactions as well. Local action in the world of infectious disease also means understanding human behavior. Understanding how disease spreads and how to control disease requires a detailed understanding of human culture. Achieving such an understanding requires us to observe how people in other cultures behave, to empathize with them, and to find ways to work with them to understand disease. René Dubos was explicit about this:

> It is a dangerous error to believe that disease and suffering can be wiped out altogether by raising still further the standards of living, increasing our mastery of the environment, and developing new therapeutic procedures. The less pleasant reality is that, since the world is ever changing, each period and each type of civilization will continue to have its burden of diseases created by unavoidable failure of biological and social adaptation to counter new environmental threats.

These evolutionary, ecological, and cultural elements place human beings at the center of the discussion of infectious disease—and therefore at the center of the discussion in this book. Because the human population is part of a vastly larger picture, in this book we describe the natural history of infectious disease in terms of expanding concentric "rings" radiating outward from the human center to encompass microbes and their interactions with humans. We first detail the human position in the world of infectious disease by explaining **ecological, cultural, and evolutionary issues** germane to human-microbe interactions (section 1). We then describe characteristics of microbes, both pathogenic and benign, that are important factors governing human-microbe interactions as humans are first **exposed** to microbes (section 2). We next move out from potential interactions, to actual interactions in the form of **infection** (section 3). Once microbes have infected individual humans, the circle widens to include how microbes travel between humans during **outbreaks** (section 4). In section 5, the circle widens even more as we consider the establishment of disease throughout populations during **epidemics**, and globally in **pandemics**. Section 6 highlights René Dubos' famous phrase, "think globally, act locally," suggesting that humans can shrink the global circle of infectious disease through effective local **action**

Section One: Evolution, Ecology, and Culture

Computer-generated image of dysentery-causing bacterium (*Shigella dysenteriae*).

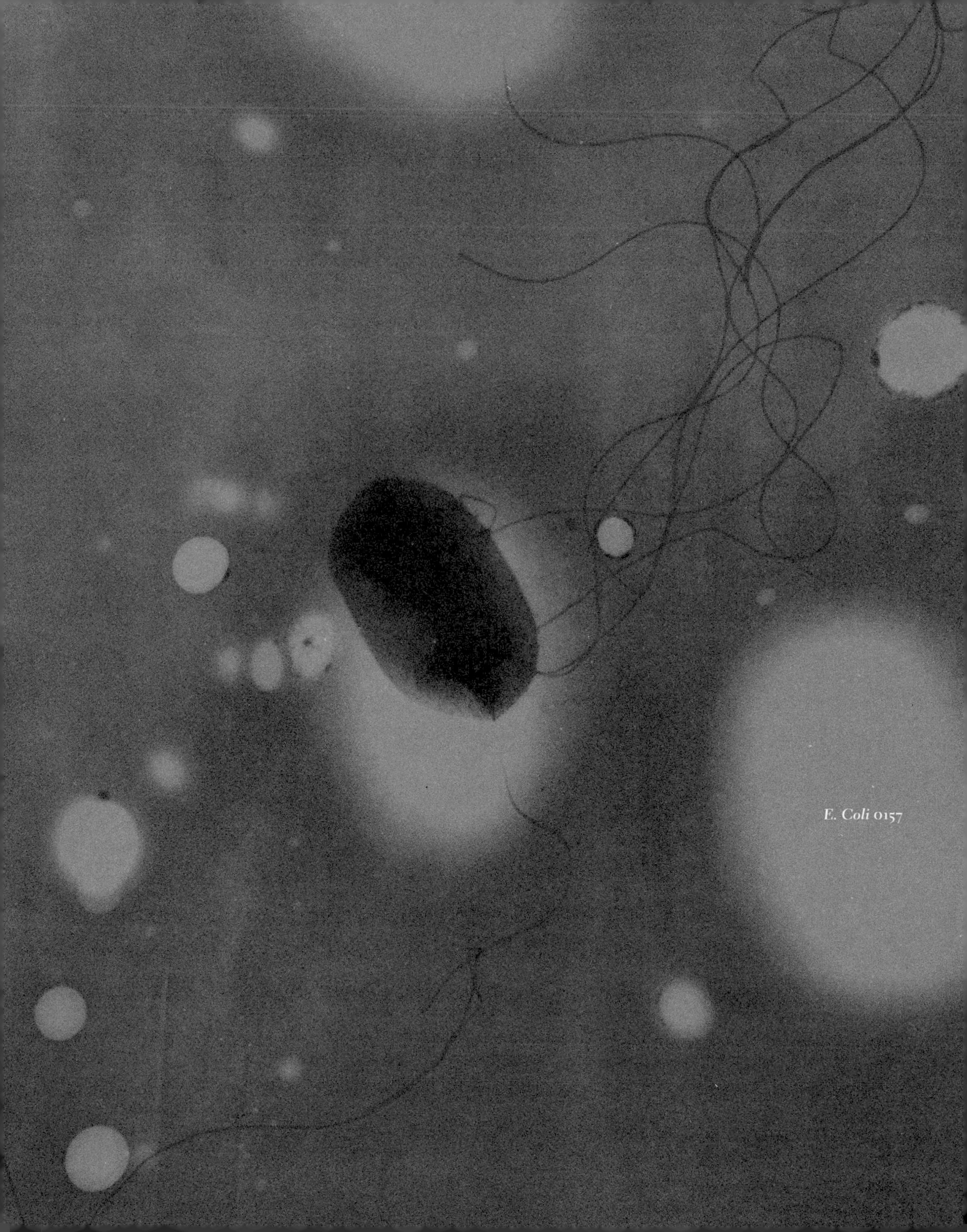

E. Coli 0157

Section One: Evolution, Ecology, and Culture

Introduction ROB DESALLE

Humans are at the center of all issues related to infectious disease. There are three keys to understanding the role of infectious disease through a natural history perspective: evolution, ecology, and culture. The first key is realizing that infectious disease occurs in an evolutionary context in which all of the Earth's organisms are bound by the same rules. The major evolutionary rule governing all organisms is natural selection. This overriding force in nature acts as a selection device—a sieve, so to speak—by ensuring that organisms that reproduce better than others end up "inheriting the earth." Here, "reproduce better" means reproduce more efficiently, which would result in more surviving offspring. The more efficient organisms are at reproduction, the more successful they will be in the evolutionary sense. Consequently, populations of organisms are under intense pressure to "adapt" to their surroundings so that they can leave more offspring.

Classical experiments on bacterial growth in vats called chemostats show the sheer power of natural selection. Varieties of bacteria of the same species can be mixed in a single chemostat, and environmental conditions (usually food source or temperature) can be varied to stimulate the evolution of the bacterial population. Some starting bacteria in the mix can be of a mutant strain that processes a particular food source more efficiently than normal bacteria do. If the mutant "handles" the food source in the chemostat better than the normal bacteria, then each individual mutant bacterium will grow faster and therefore reproduce at a younger age—say fifteen minutes instead of twenty minutes—than will each individual normal bacterium. After enough time has elapsed, the total bacterial population will have greatly increased and the vast majority of bacteria in the chemostat will belong to the mutant population. In evolutionary terms, the mutant population has been "selected" over the normal population by the environmental conditions in the chemostat.

As suggested by this example, natural selection decides which genes in a population of a particular species are passed on from generation to generation. However, natural selection is only indirectly involved in any battle between species. In bacteria-caused infectious disease in humans, human interventions such as natural immunity (honed through biological evolution, driven by natural selection) or the use of antibiotics (honed through cultural evolution) are selective agents that directly affect the survival of the bacteria. The efficiency with which bacteria manage to infect people and the virulence of the resulting infection directly affects the survivability and health of people. Thus, when different species interact, natural selection is a factor that decides which species does better, even if it is an indirect one.

The second key concerns an understanding of ecology and ecosystems. The classical definition of an ecosystem is "a bounded system where living organisms coexist in a constant state of interaction and interdependency." The human body is the habitat of a community of microbes that mutually interact; in this way the body is an ecosystem. Indeed, when microbes interact with humans, they typically remain in specific parts of the human body—the bloodstream or the respiratory tract—and thus are in a "bounded system." Many species of microbes living within or on our bodies are necessary for our survival. These microorganisms, our "normal flora," have evolved to live in specific parts of our bodies. In taking advantage of our hospitality, most of them do us no harm, and some benefit us. For instance, many strains of bacteria belonging to the species *E. coli* have evolved a mutually beneficial relationship with human digestive tracts. There they find nourishment while making it difficult for other harmful bacteria to settle. *E. coli* can survive only in an animal's gut and can exist only ephemerally in places outside animal digestive tracts (except for those cultured in the laboratory). *E. coli* has not evolved to have beneficial relationships with other parts of our bodies, therefore, if it begins to grow in areas other than our digestive tracts, it becomes a dangerous, and potentially deadly, pathogen. The human body ecosystem is not always critical to every pathogen that infects humans. For instance, the ecosystem of pathogenic microbes that interact with humans through other vector species—for example, a mosquito or tick—are affected by the vector's ecosystem, as well as the human body.

Within a given ecosystem, the "constant state of interaction and interdependency" among all of its living species has obvious evolutionary implications. The force of natural selection that drives the evolution of each of the ecosystem's species derives largely from all the other species present. In other words, each species in an ecosystem directly or indirectly drives the evolution of every other species. As populations of different species coevolve in the same evolving ecosystem, the nature of the relationship between any two species cohabiting—for example, between a pathogen and its host—will continuously change, sometimes to the advantage of one species over the other. Ecologists debate how much change typically occurs in a healthy ecosystem, and about how much the mix of its different species can change before it becomes "unhealthy" or "out of balance." However, there is much evidence that the loss of certain key species in an ecosystem—perhaps through drought, fire, deforestation, introduction of alien species, etc.—can drastically affect all the remaining interactions in the system. Host–pathogen interactions are among those that can strikingly change when an ecosystem is "assaulted" in such a manner.

The final key is culture. Different cultures define "disease" differently. What Western culture regards as a disease might be regarded otherwise in non-Western cultures. According to the social scientist Robert Hahn, "How we think of sickness and the different kinds of sickness shapes our response, diagnosis and treatment." For instance, some cultures regard certain microbe-caused afflictions as part of daily life—or even as significant milestones in life—rather than as "disease." Measles debilitates many humans across the globe, and in most cultures is regarded as a "disease" because it produces discomfort and sometimes death. However in Hong Kong, China, a traditional belief holds that the effects of the measles virus is simply a transition—from one stage of life to another. Infection by measles is a rite of passage from childhood to adulthood and is not perceived as a disease.

Western culture has generally accepted a biomedical perspective on disease, which guides our definition of "disease." Westerners assume that the medical approach to infectious disease is rational, systematic, and scientific. However, this does not mean that non-Western perspectives on disease are inadequate or naive. In fact, some non-Western cultures have extremely sophisticated perspectives on infectious disease, with detailed methods and terminology for diagnosis, treatment, and disease prevention—even if these do not derive from a biomedical perspective.

To explore evolution, ecology and culture, I pose the following questions:

Given the number, the diversity, and the nature of microbes that cause disease, why are we still here?

JOSHUA LEDERBERG, President Emeritus of Rockefeller University and Sackler Foundation Scholar, answers this question by describing human and microbial interaction in an evolutionary context.

Where do new diseases come from?

ANDREW SPIELMAN, Professor of Tropical Public Health at the Harvard School of Public Health and head of the Laboratory of Public Health Entomology, explains the dynamics of emerging infectious diseases and how changing conditions may disrupt the existing balance between humans and microbes.

How does biodiversity affect disease?

FRANCESCA T. GRIFO, Director of the Center for Biodiversity and Conservation at the American Museum of Natural History, explores this question by examining how deforestation alters disease patterns.

Through the careful exploration of these questions, the authors of the essays introduce how evolution, ecology, and culture interact to create a shared ecological system—and how human interpretations of that system structure societal response to infectious disease.

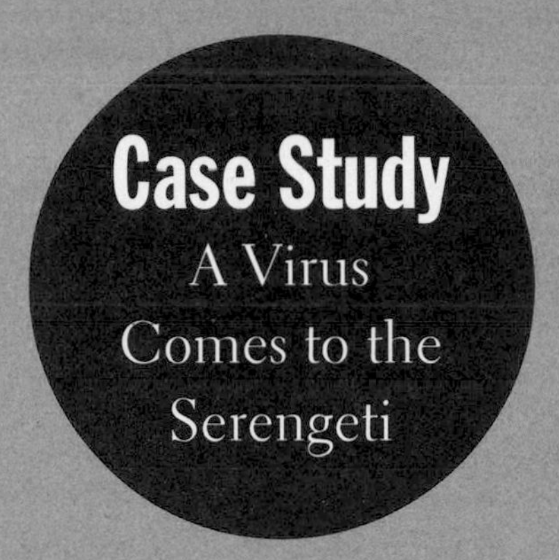

Throughout the book, a series of case studies looks at the process of scientific discovery by closely examining disease outbreaks and how we fight them.

Until the last decade of the nineteenth century, the Serengeti Plain, in the East African nation of Tanzania, was an Eden. Many species of animals and plants existed side by side, each contributing to the ecosystem, a splendid balance in which humans and even microscopic disease-causing organisms had a place. But in 1890 something happened to tip that balance, causing a cascade of events that disturbed the biodiversity and affected every living thing on the plain.

Plant life in the Serengeti consists mostly of short and long grasses, punctuated by thickets of several varieties of the acacia tree. The grassy prairie is ideally suited to a wide assortment of wild grazing ungulates (hoofed animals), including the blue wildebeest, the African buffalo, the plains zebra, several species of antelopes, among them the dikdik, the Thomson's gazelle, and the Grant's gazelle, and browsers such as the giraffe. In addition to these wild ungulates, the Masai and Sukuma peoples who lived in settlements bordering the plain introduced domesticated varieties of the now extinct auroch. In this stable ecosystem, the ungulate herbivores thrived, their populations limited solely by food resources and by their natural predators: lions, cheetahs, leopards, hyenas, and jackals.

Two much smaller organisms coexisted with these animals and plants and to a large degree prevented a large-scale human presence, which would upset the delicate balance. The tsetse fly, which found shade and suitable mating and breeding habitats in the acacia thickets, is host to the even smaller *Trypanasoma rhodescens*, a protozoan that is harmless to wild ungulates, but causes a fatal nerve disease in humans and domesticated cattle.

The tsetse feeds on the blood of ungulates, which are unharmed by the bite. The tsetse acts as the vector of *T. rhodescens*, repeatedly spreading the protozoan throughout the ungulate population while taking its blood meal. The ungulates provide a reservoir for *T. rhodescens*, their bloodstreams a safe place for it to reproduce.

When the tsetse bites a human, however, the results are deadly. Once bitten, an untreated human dies within three to twelve months from neurological damage. In its final stages, the illness causes extreme drowsiness, which accounts for its common name: sleeping sickness.

Except for the relatively small numbers of Masai and Sukuma people, humans were unknown on the Serengeti, and the well-balanced ecosystem survived in its edenic state. Around 1890, however, the balance was drastically disrupted with the arrival of Europeans and the creatures large and small they brought along with them.

It is uncertain whether it was the Italians or the British, but the purpose was conquest. The British came to East Africa in 1884, hoping to maintain control of what is now Sudan. The Italians came in 1889, attempting to invade Ethiopia. Both brought along cattle, and some of the cattle were infected with a virus endemic to the Black Sea Region of Asia, but previously unknown in Africa.

The rinderpest virus (RPV) is highly infectious and, unlike *T. rhodescens*, did not spare the wild ungulates of the Serengeti. An infected animal typically died within three weeks. Two years after the introduction of RPV into East Africa, most of the domestic cattle were dead. By 1896, there were but a few wild ungulates on the plain.

The loss of both wild and domesticated ungulates had a devastating effect on the Masai, whose food supply depended on both their own herds and the wild ungulates they hunted. The Sukuma, who engaged in subsistence farming as well as keeping cattle, were slightly better off—that is, until the Masai began raiding their cattle, which served also to further spread RPV.

The degree of starvation was so great that "People were reduced to chewing the bark of trees," according to a contemporary report. Weakened by starvation, the people of the plain were more vulnerable to infectious diseases like smallpox, which swept through the population, reducing their numbers and making it all the more difficult for them to maintain or obtain food supplies.

Without the ungulates to provide the blood required by the tsetse, the flies disappeared, and with them *T. rhodescens*. The lions also lost their natural prey when ungulate numbers declined and they shifted their attention to humans. Fearing attacks by man-eating lions, the remaining Masai and Sukuma abandoned their villages and fled the area.

Like a line of dominoes toppled by the tipping of a single tile, the Serengeti ecosystem fell into disarray. But natural systems adapt to new conditions in an attempt to regain balance. That's what happened in the Serengeti, beginning around 1910.

With the abandonment of the villages and of the farming associated with them, combined with the greatly reduced numbers of grazing and browsing ungulates, grasses and trees flourished. In time, the small numbers of remaining ungulates—clearly the strongest—increased, and many of them had developed resistance to RPV. The tsetse came back to the thickets, and with them came *T. rhodescens* and the sleeping sickness so fatal to humans. This further discouraged the Masai and Sukuma from resettling the areas bordering the plain.

RPV resistance was not permanent, however, because the virus is able to mutate in ways that evade the immunity developed by the ungulates. Approximately once a decade between 1917 and 1959, outbreaks of RPV caused several populations of buffalo, wildebeest, giraffe, and Masai cattle to die off.

Beginning in the 1930s, human intervention once again played a part in the Serengeti ecosystem. Mechanical clearing of the bush helped hold the tsetse population in check, encouraging the Masai and Sukuma to return. Then, in 1952, a program to vaccinate domestic cattle against RPV protected not only the Masai and Sukuma herds, but also prevented its spread among wild ungulates. By the early 1960s, RPV had died out on the Serengeti Plain.

Although RPV is no longer a problem, sleeping sickness persists, though it is less of a threat than it was during the first half of this century. Drugs are now available that can prevent or cure it, if taken in time. Eradicating the tsetse entirely is a daunting challenge, not least because it would require eliminating its habitat, the acacia thickets, a fundamental component of the ecosystem that is now the Serengeti National Park, as close to Eden as the twentieth century allows.

There are many lessons to be learned from the Serengeti story. One of them is that a disease that is not infectious to humans can nonetheless have a profound effect on them. By disrupting the balance in an ecosystem, it can destroy their food resources and lead to conditions favorable to other diseases that do infect humans, often fatally.

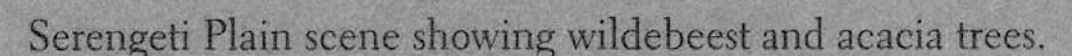

Serengeti Plain scene showing wildebeest and acacia trees.

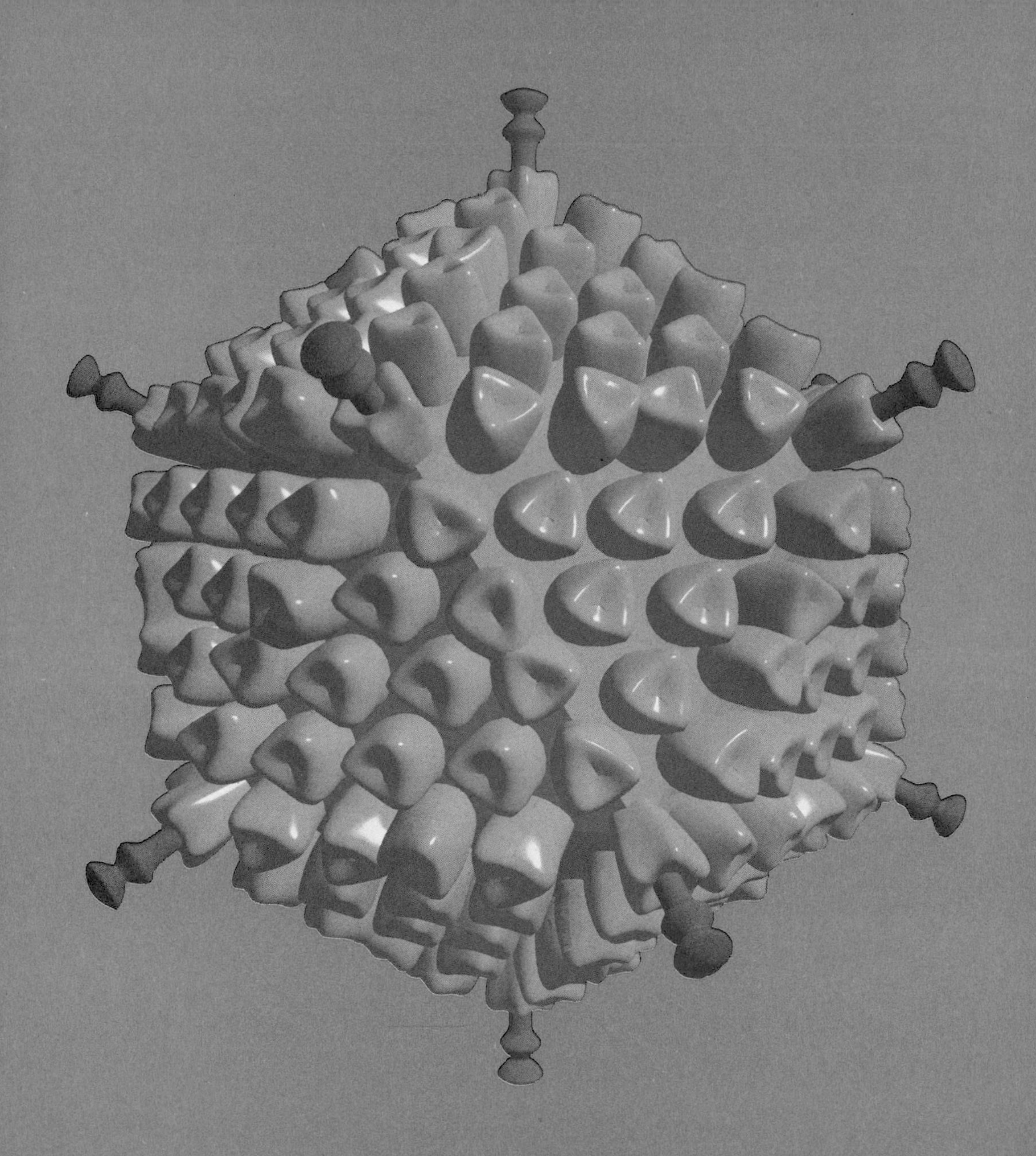

Computer-generated image of a common cold virus (Adenovirus).

Infectious Disease as an Example of Evolution

JOSHUA LEDERBERG

Let us look at our relationship to infectious pathogens as part of the evolutionary drama. Here we are; here are the bugs. They are looking for food; we are their meat in one sense or another. How do we compete with them? There are so many of them, they reproduce so much more quickly than we do, they tolerate vast fluctuations of population size as part of their natural history. They lack the emotional apparatus to grieve whenever their population significantly declines. We are different from them in every respect. We regard a reduction of one percent in our population size as a major catastrophe. One hardly need say more than that.

The sheer number of microbes, the rapidity with which they can reproduce, and their tendency to change their genetic makeup gives them an enormous potential when it comes to acquiring advantageous adaptations. The bugs have the potential to make toxins, to vary, to find defense mechanisms against our antibiotics, and to alter their proteins such that they escape detection by our immune system.

In pondering all this—and I have spent all my life looking at microbial variability and diversity—I am left with the question, why are we still here? It's perfectly easy to imagine the microbe that could wipe us out. We have had some close calls with real microbes. For example, the Spanish influenza pandemic of 1918 killed twenty to twenty-five million people, about 0.5 percent of the world's population. The toll of the fourteenth-century plague, the "Black Death," was closer to one third. If the bugs' potential to develop adaptations that could kill us off were the whole story, we would not be here. However, with very rare exceptions, our microbial adversaries have a shared interest in our survival. Almost any pathogen comes to a dead end when we die; it first has to communicate itself to another host in order to survive. So historically, the really severe host-pathogen interactions have resulted in a wipeout of both host and pathogen. We humans are still here because, so far, the pathogens that have attacked us have willy-nilly had an interest in our survival. This is a very delicate balance, and it is easily disturbed, often in the wake of large-scale ecological upsets.

Joshua Lederberg is a Sackler Foundation Scholar and President Emeritus of The Rockefeller University in New York.

Microbes—here I can group together bacteria and viruses—do not all have the same adaptations for attacking their hosts; however, the same principles apply no matter which group of microbes we are referring to. They all multiply rapidly and constantly; they have huge population sizes—the numbers used to indicate these sizes end in twelve, thirteen, or fourteen zeros. The time between generations is only minutes.

Even when individual microbes reproduce merely asexually, just replicating their own genetic material and then dividing into two, they often produce genetic variants of themselves. Processes that help maintain precise replication can be "turned off" because of DNA (dioxyribonucleic acid) damage or other injury. Compared to ourselves, bugs often live in a sea of mutagens. RNA (ribonucleic acid) replication is particularly error prone. Unlike the case with DNA replication, there are no editing mechanisms to correct errors that occur during RNA replication. For many RNA viruses, the rates of mutation are so high that to a close approximation, every virion is genetically different in at least one nucleotide position from every other one. Therefore, no single genotype is totally representative of all the genotypes present in this swarm. Each swarm of genotypes is rapidly evolving through the process of natural selection because some genotypes within the swarm express themselves in the form of more effective adaptations to the environment than those expressed by most of the other genotypes.

There are also genetic factors controlling the rates of mutation; whether these affect pathogens' virulence is a matter of controversy right now. Some preliminary reports had suggested that virulent bacteria had a higher incidence of mutations than did nonvirulent bacteria. We now realize that mutations are quite prevalent and therefore bacteria are facing environmental changes all the time whether they are pathogens or not. What would be of particular interest—and there is beginning to be increasing evidence for this—is if even the tendency to mutate were physiologically regulated by external stress. Mutation rate is not a constant of physics; it is a variable of nature, and bacteria will be more mutable if they are subjected to nutritional stress than if they are living in a perfectly stable environment where all they need or want to do is multiply the same way they have been all along.

To cap all this, many bacteria and viruses often have the capacity to exchange genetic information. This may occur through some primitive forms of cellular sexuality, or by the interaction of DNA molecules from different species, migrating through particles we call plasmids or viruses. We can think of the microbial world as an interconnected World Wide Web of genetic information.

Given the enormous capacity for genetic plasticity on the part of microbes, evolutionary adaptation is not a serious limitation for them. It is their shared interest in our survival that will dominate the overall picture of their (and our) evolution. There is a long-term trend of evolution by both microbe and host through which the host species acquires factors for resistance and the parasite species becomes less virulent to the host species. This allows the host species to survive longer after infection, thus, simultaneously lengthening the time the parasite can survive in the host.

For their part, the pathogenic species will find it to their advantage to evolve in a way that reduces their virulence, provided they can do so without compromising their livelihood. And that is the tightrope that they walk. The bug that lets its host live for another day also survives another day, allowing its own propagation and its own spread to new hosts as well. We have enough examples of apparently fairly stable coadaptations of pathogen and host that we can rely on this being a frequent, but not inevitable, outcome in these long-term relationships. I will take the rhinovirus, the microbial cause of the common cold, as an example of an extremely successful pathogen. We do not do much to try to get rid of it; we go to work, school, and play, with our runny noses. The rhinovirus has several adaptations, including the very moderation of its disease process that tend to facilitate its spread. There are any number of varieties of it. One could hardly ask for more.

A pathogen-host coadaptation of this type is certainly marked by short-term flare-ups. Even if you have a reasonably stable situation, any member of the pathogen population that had a mutation that enabled it to relax its constraints would reproduce faster than its neighbors, cause great harm to its host, and act very much like a cancer cell does within the "properly restrained" cell population making up a human. Something like that seems to be happening with HIV infection. Normally, the body can be infected with HIV for a fairly long period of time, during which there are many cycles of coadaptation, if you like, between the evolving population of HIV viruses and the body's population of antibodies to these viruses. Eventually, however, some further mutation in the HIV virus occurs, or the wear and tear on the most vulnerable component of the body's immunological system, a particular type of white blood cell that stimulates most of the other infection-fighting cells into action, becomes too severe. The host-parasite balance is then tipped such that the host becomes severely ill and dies, not of the HIV infection but of other diseases, which would not affect people with unimpaired immune systems. And that is the doom facing the population of people infected with HIV at present.

So there is a parasite's dilemma. If it proliferates rapidly within a host, it may kill the host as a byproduct. However, such a rapid proliferation would be a winning strategy if the parasite were easily transmitted to a new host. This would be the case if vectors are ready at hand and we have obliging behavior by the host carrier. We see this with *Plasmodium falciparum*, one of the four microbe species that cause malaria. *Plasmodium falciparum* would be unlikely to survive for long because of its high virulence. The high density of mosquito vectors of *Plasmodium falciparum*, enable it to spread rapidly and reliably from infected people to uninfected people.

There is an inexhaustible reservoir of potent microbial toxins, and one wonders why they have not spread much further. Botulinum toxin, one of the deadliest chemical compounds known, is produced in abundance by the bacterial species *Clostridium botulinum*. It's easy to imagine that the genes that govern the toxin's production could spread to other organisms, thereby causing botulism to become one of our major public health threats. Why does the capability of producing this toxin remain confined to *Clostridium botulinum*? It is not confined by any underlying genetic mechanisms. Rather, any new microbes that deployed this weapon would likely kill off their host so rapidly that they would most likely die themselves before they could infect another host, unless they shared *Clostridium botulinum*'s unusual ability to multiply in a dead host, in combination with the ever-looming saprophytes: the agents of putrefaction. Consequently, most transfers of the genes that would produce this toxin will be evolutionary dead ends.

Suppose a pathogen adopts the strategy of proliferating slowly in, say, a human host. Such a pathogen has problems because within a week,

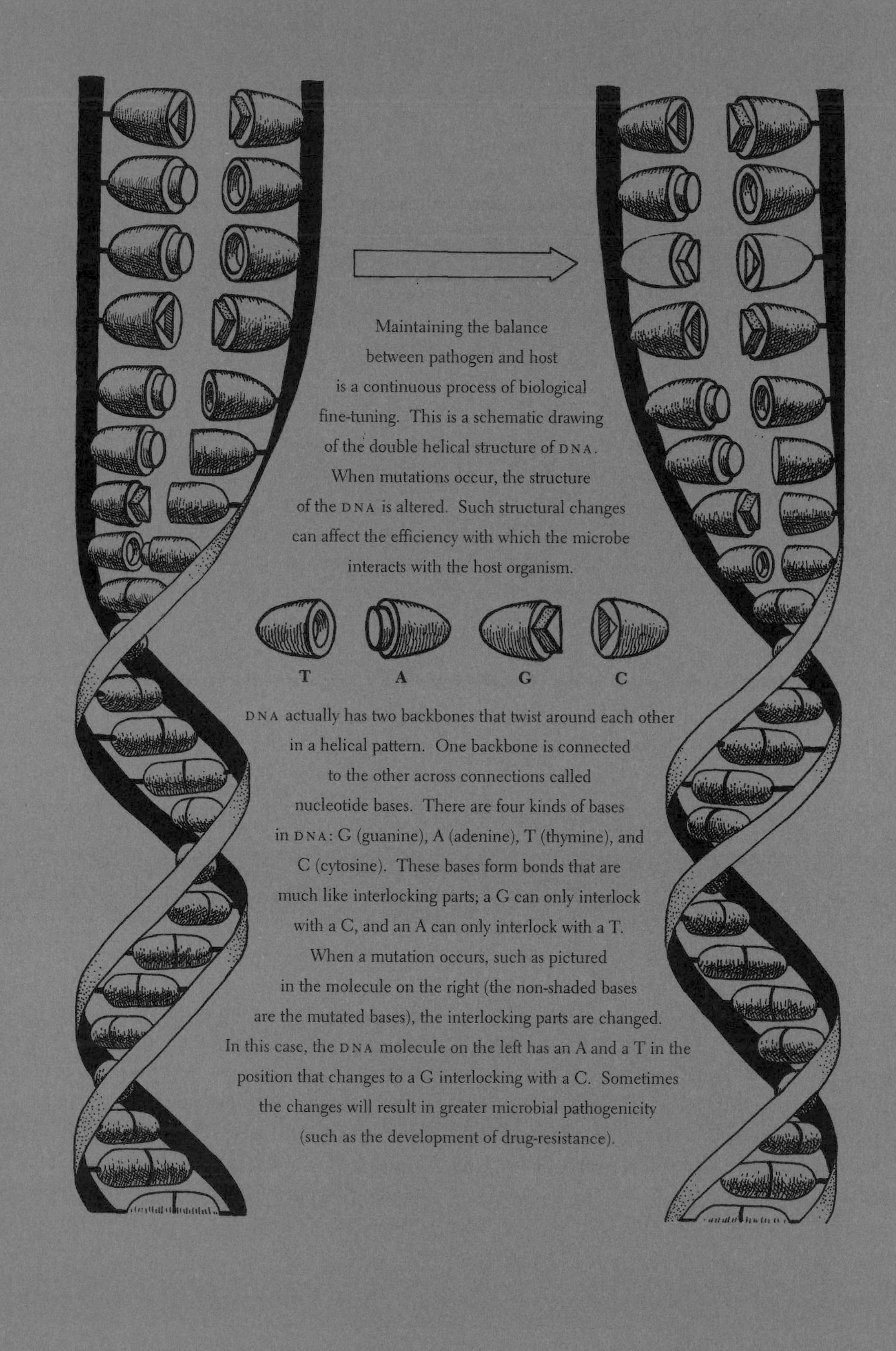

Maintaining the balance between pathogen and host is a continuous process of biological fine-tuning. This is a schematic drawing of the double helical structure of DNA. When mutations occur, the structure of the DNA is altered. Such structural changes can affect the efficiency with which the microbe interacts with the host organism.

DNA actually has two backbones that twist around each other in a helical pattern. One backbone is connected to the other across connections called nucleotide bases. There are four kinds of bases in DNA: G (guanine), A (adenine), T (thymine), and C (cytosine). These bases form bonds that are much like interlocking parts; a G can only interlock with a C, and an A can only interlock with a T. When a mutation occurs, such as pictured in the molecule on the right (the non-shaded bases are the mutated bases), the interlocking parts are changed. In this case, the DNA molecule on the left has an A and a T in the position that changes to a G interlocking with a C. Sometimes the changes will result in greater microbial pathogenicity (such as the development of drug-resistance).

more or less, the host's immune system will have produced antibodies that cause the pathogens to be recognized and destroyed by the host's defenses. So the bug that proliferates slowly is highly vulnerable unless it adopts some further tactics. Such tactics could include coating itself with proteins that are similar enough to some of the host's proteins that the host's immune system fails to recognize that these pathogens are "foreign invaders" and, therefore, ignores them as if they were part of the host's own tissue. We should also keep in mind that parasites are competing with a variety of microbial species that normally live within the host, where they may actually benefit the host and, at least, do not harm it under normal circumstances. Here is where the HIV virus runs into severe trouble. Left to its own devices, HIV wouldn't kill its host. However, by knocking down the immune system of the host, it leaves the door open for other organisms to thrive, including ones that commonly inhabit the host without harming it, as so-called opportunistic infections, the end result being the death of the host. Other mild infections may be more successful by enhancing the host's defenses against other pathogens—we sometimes see evidence of this as a side effect of antibiotic treatment.

If we follow this line of thinking a little further, we are not surprised to observe that vectors rarely show any symptoms, or at least do not show severe symptoms. For example, it would not do *Plasmodium* very much good if it killed the mosquitoes that were on the way to transmitting it to humans. We see this pattern over and over again. We do not usually refer to a "mad" or rabid dog as a "vector" for the rabies virus. However, the unusual behavior of a rabid dog—its greatly increased tendency to bite humans and other animals—is a behavioral adaptation that "serves the virus's purpose" by spreading it from one of its victims to another.

But the ultimate symptom—the death of the host—is almost never to the advantage of the parasite. Whenever we see that happen, we have to regard it as a breakdown in the equilibrium, in the contract, in the arms-control agreement, if you like, between pathogen and host. The outcome could have been better had both sides been more witting.

The implication of this train of thought is that we should consider symptoms from an evolutionary perspective: what was their origin? Asking this type of question may open the door to new avenues of thought in examining the disease process.

Adapted from Opening General Session Address from the International Conference on Emerging Infectious Diseases, Atlanta, GA, March 8, 1998.

How has man influenced the evolution of infectious disease?

How can the course of infection be influenced by the presence of a single genetic mutation?

Why has the presence of various toxins had little effect on the evolution of infectious diseases?

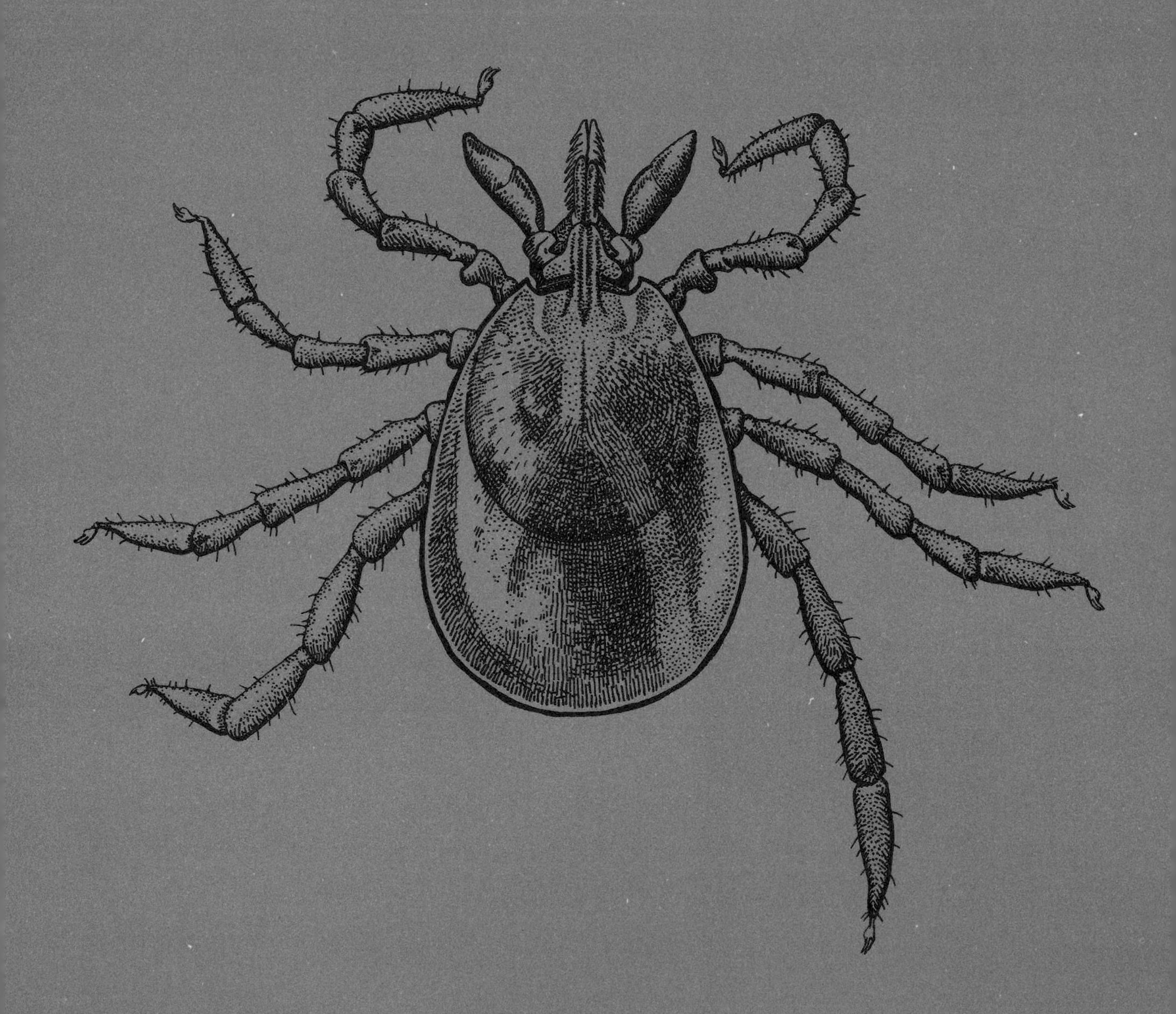

Deer tick, *Ixodes daminni*.

Emergence of New Diseases

ANDREW SPIELMAN

INFECTION IS EVERYWHERE AND GENERALLY "SILENT"

Microbes are everywhere. They make up about a twentieth of our weight, filling our intestines, covering our skin, and lining our mouths and nasal passages. Microbes exist in our food and drinking water. A suitable microbial mix is essential to our well being. If certain of these microbes that have colonized us are destroyed, as by excessive antibiotic use, our microbial balance may be upset, which can result in such problems as diarrhea, vaginitis, bad breath, or body odor. More serious problems may follow. Human beings, like other animals, cannot function without their microbial companions.

The outcome of the host-parasite relationship between humans and microbes, of course, may also be devastating to humans. Three examples are: bubonic plague, typhus, and malaria. The "great plague" epidemics of the Middle Ages decimated Europe. This bacterial infection of rats and humans generally infects new hosts through the bite of a flea that has bitten an infected rat. When it is airborne, plague can also be carried directly from person to person via sneezes and coughs. In such cases, it infects the lungs or the blood, often with fatal consequences. Similarly, epidemic typhus can destroy entire communities. Refugee populations are particularly vulnerable to this louse-transmitted disease because they often lack access to adequate hygiene and clean water. More than a million people still die of malaria each year.

We encounter many microbes that infect our respiratory and gastrointestinal systems each year, though not all are lethal or debilitating. Many of these infections are "silent"; that is, they produce no symptoms. Others merely cause a transient fever or a brief period of indigestion. Some pathogens, like the viruses that cause measles, mumps, and chicken pox, affect specific sites of the body. Such highly prevalent infections are usually most damaging when people first have them after reaching adulthood. Polio is then more likely to result in paralysis; mumps, in sterilizing bouts of testicular swelling; chicken pox, in painful episodes of shingles; and measles, in encephalitis.

Our bodies "remember" each episode of infection and accumulate a set of immunological agents

Andrew Spielman is Professor of Tropical Public Health at the Harvard School of Public Health and Principal Investigator of the Laboratory of Public Health Entomology.

that prevent subsequent infection of the same pathogen. Moreover, vaccines are available that protect us from an initial infection by many of the more common pathogens. Certain microbes, such as those that cause AIDS, inhibit our ability to cope with pathogens.

In sum, the microorganisms that invade our bodies become part of our daily lives. They are many. We adapt to their presence, and they adapt to us. The following discussion explores the rules that govern this interaction and explains how changing conditions may lead to the development of novel disease relationships. New diseases emerge when people come into contact with microbes associated with other animal hosts, either human or non-human.

INFECTIONS PROTECT NON-MIGRATORY HUMAN POPULATIONS FROM POTENTIAL INVADERS

Malaria nearly defeated Scottish explorer Mungo Park's attempt to discover the headwaters of the Niger River in West Africa. It contributed to his eventual ambush and drowning in 1805—by killing all but four of his forty-four European companions. His expedition suffered continuously, apparently mainly from malaria. Many of its members had to be carried by friendly residents of the region, who apparently enjoyed robust good health. During 1854–59, the American explorer Paul DuChailu survived more than fifty "episodes of fever" in the same region by dosing himself regularly and liberally with quinine; malaria incapacitated him for nearly a year.

This part of West Africa, which included the infamous Bight of Benin (the bay off the "Slave Coast"), was the "white man's grave" of the early 1800s. More than a tenth of the Europeans who attempted to live there died within a year after their arrival, mainly from malaria. In contrast, few natives of the region seemed to suffer from this infection. West African warriors, therefore, enjoyed an overwhelming advantage over their potential foreign invaders. Because this mosquito-borne infection is "rooted in the village environment" where local residents "learn" to tolerate this infection in infancy, malaria protects them by eliminating foreign invaders.

Europeans could colonize West Africa only after devising effective interventions against malaria. Reflecting the colonial ethic of early twentieth-century Germany, the great microbiologist Robert Koch wrote, "We will not be happy in our colonial possession until we succeed in becoming master of this disease." A half century later, after decolonization, a West African scientist proposed placing a picture of a mosquito on the Nigerian flag because it "saved the land of our fathers for us." One's enemy's enemy may be one's friend.

OTHER INFECTIONS HELP INVADERS DISPLACE NON-MIGRATORY HUMAN POPULATIONS

Some infections favor migratory populations at the expense of non-migratory populations they may contact. This typically happens when the migratory population is large enough to maintain infections that produce lifelong immunity, while the non-migratory population is too small to sustain the infection—and thus has lost any resistance it might have had to the infection-causing pathogen. Measles is one of the most communicable of infections transmitted person-to-person. The supply of non-immune human hosts that "fuel" successive measles outbreaks in a limited area is replenished slowly, mainly through births or immigration. When fewer than one third of a million people reside in a given area, this accumulation of susceptible hosts is so slow and unreliable that measles is likely to die out. Mumps, which is less communicable, can perpetuate in smaller populations—as can polio, chicken pox, and German measles.

Although such directly transmitted "infections of childhood" as measles, mumps, chicken pox were rarely life threatening in nineteenth-century London, Paris, or Berlin, they devastated the Eskimo and American Indian villagers when they came in contact with infected Europeans. Affected native adults suffered paralysis, orchitis (inflammation of the testes), shingles, and encephalitis. Children were born with severe birth defects. The native Wampanoag Indians of Nantucket Island (off the Massachusetts coast) were much larger and seemingly healthier than their European visitors at the time of initial contact. Nevertheless, they soon succumbed to the "indian sickness," leaving their lands to their foreign guests. The specific diseases afflicting these natives may have included louse-borne typhus and relapsing fever. These spectacularly healthy original inhabitants of the region might have fared better if they harbored a microbial enemy for their potential enemy.

OTHER ANIMALS RELY ON MICROBES TO DISPLACE OTHER POPULATIONS

White-tailed deer act as a reservoir for pathogens that infect other animals. Surprisingly, they benefit from the effects of their pathogens on other animals. As animals succumb to infectious diseases, there is less competition for their food supply. The meningeal worm is an apparently benign nematode (roundworm) that resides in the cranial meninges (the covering between the brain and the cranial bones) of these deer. Its larvae pass in the deer's feces. The worm's life cycle is completed when another deer eats a terrestrial snail or slug that has ingested these larvae. However, if elk, moose, or other kinds of deer ingest the larvae of these worms, they can sustain severe neurological disease.

Two other species also affect these animals much more severely than they do white-tailed deer. The arterial worm, *Elaeophora schneideri*, develops in the (neck- and brain-supplying) carotid arteries of white-tailed deer and spreads through bites by deer flies and horse flies. The abdominal worm, *Setaria yehi*, infects the abdominal cavity of these deer and spreads by bites by *Aedes* mosquitoes. Similarly, the liver fluke *Fasciola magna*, a flatworm, causes no symptoms in white-tailed deer but is generally fatal in sheep or cattle. White-tailed deer also serve as the main reservoir hosts for various pathogens that cause human disease. These pathogens include the mosquito-borne Jamestown Canyon and Cache Valley viruses and the tick-borne agent of a bacterial pathogen, *Ehrlichia chareensis*. The list of pathogens that white-tailed deer may share with their neighbors is long.

THE EXAMPLE OF EMERGENT LYME DISEASE

White-tailed deer are indirectly associated with a surprisingly broad array of microbes that exploit the white-footed mouse as a reservoir host. These microbes, which are carried to people through the bites of ticks that have previously bitten infected deer, include the following disease-causing agents: the bacterium *Borrelia burgdorferi*, which produces Lyme disease; the protozoan *Babesis microti*, which produces human babesiosis; and the bacterium *Ehrlichia microti*, which produces human granulocytic ehrlichiosis. White-tailed deer are crucially involved in the transmission of these infections because the adult tick typically feeds on them. The deer tick, *Ixodes dammini*, that transmits these pathogens is abundant in certain regions in the northeastern United States. A related species of tick infests more southerly states. These diseases become common only where deer are abundant.

The early European visitors to eastern North America described a heavenly landscape. Groves of stately trees stood among vast meadows. Brush and broken trees were absent because the original inhabitants of the region shaped the land with fire. They harvested only as much game as they intended to eat. The situation soon changed. When Thoreau lived by the shore of Walden Pond in the 1830s, he wrote of the destruction of the original forest and the total absence of game. The local iron industry was demanding vast quantities of charcoal; the railroads needed ties and wood for fuel; and people used charcoal to cook and wood to heat their homes. Settlers had converted "useless" forests and wetlands into farmland. This denuded landscape began to regenerate during the mid-1800s, when intensive agriculture began to shift to the West, and our present weedy landscape started to form.

White-tailed deer are well suited to the resulting mass of crowded young trees, fallen limbs, and brush that now surrounds our homes. Sightings of these once-scarce animals are becoming commonplace. The North American deer herd has been proliferating explosively since the mid-1900s, as have their deer tick parasites. Americans first became exposed to the infections associated with these animals during the 1960s. Physicians report thousands of cases each year in a steadily increasing pattern that promises to continue.

SUMMARY

Parasites generally contribute to the well being of their natural reservoir hosts by afflicting other organisms that potentially compete with, or prey on, their main reservoir hosts. The specific form of the disease caused by these parasites evolves in a way that it spares their natural reservoir hosts while damaging their hosts' competitors or predators. Newly arrived hosts, therefore, sustain severe damage; this provides some advantage to hosts that have been in long-term contact with the pathogen. Therefore, a disease may appear in the human population when people come in contact with another host population, either human or other animal. Disease may also emerge when the circumstances of transmission are altered because of environmental change. Such an event may convert people into a population of intruding "targets" in the natural life cycle of the pathogen.

How have people been able to tolerate the ever-present barrage of microbes?

What are the factors that led to the emergence of Lyme disease?

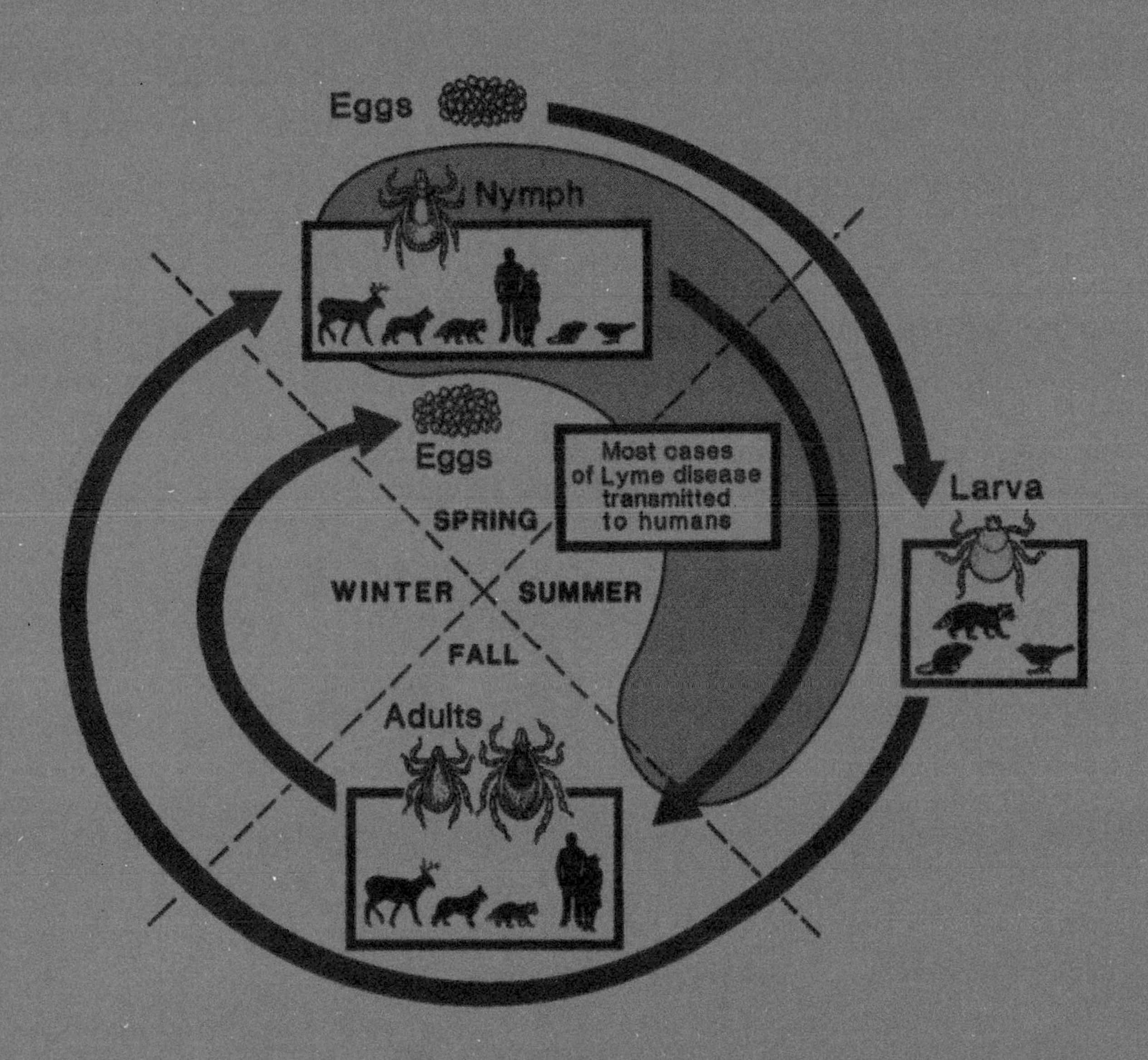

Life cycle of ticks that cause Lyme disease.

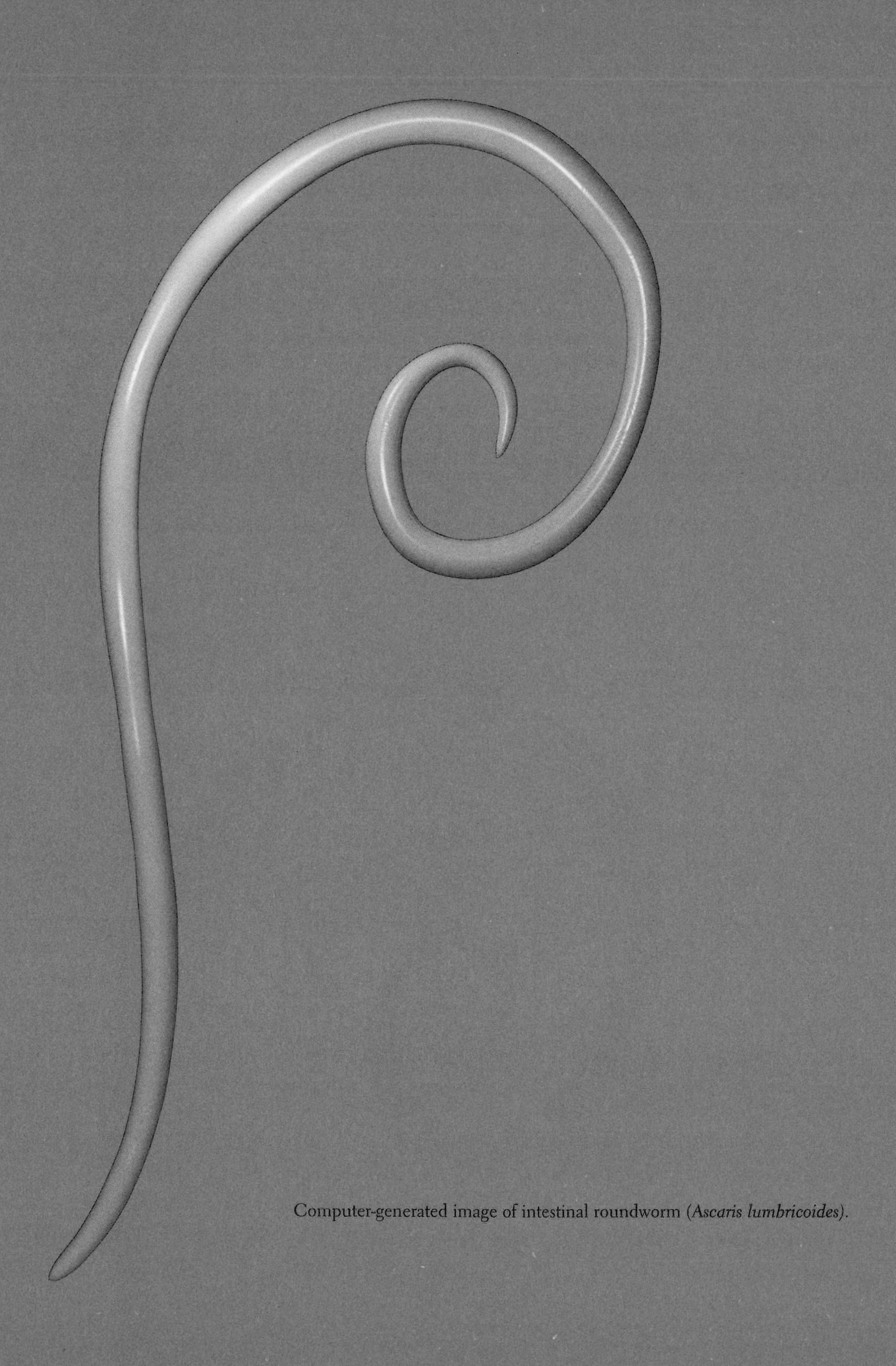

Computer-generated image of intestinal roundworm (*Ascaris lumbricoides*).

Infectious Disease and the Loss of Biodiversity through Deforestation

FRANCESCA T. GRIFO

Many of the pathogens that cause human disease are present in the environment most of the time. The reasons why we are not constantly symptomatic are varied and complex. In many instances they have a direct link to the benefits provided by biodiversity. Biodiversity, the incredible variety of living things, assures sufficient food and water supplies, keeps populations of disease-causing organisms in check, provides source materials for medical therapies, models for medical discoveries, and warnings of toxins and other environmental hazards. The disruption of ecosystems and loss of biodiversity threaten supplies of the food we eat, the water we drink, the air we breathe, and the medicines we need.

Biodiversity plays a significant role in controlling pests, pathogens, vectors, and human parasites. Disease-causing organisms, the pathogens, often have very complex life cycles in which they may utilize numerous species as hosts (places to live and or reproduce), vectors (ways to get from one place to another), or reservoirs (places to "hang out" until external conditions improve). When ecosystems are disrupted, the normal disease behavior is frequently disrupted and humans very often end up being at greater risk of becoming ill or even dying.

As we know, forest ecosystems are highly biodiverse. One of the clearest examples of how ecosystem disruption affects disease behavior can be seen in the interactions between deforestation and the infectious, and particularly the vector-borne, diseases that are common throughout the tropics and the sub-tropics.

DEFORESTATION ALTERS DISEASE ECOLOGY

People have always cut forests for two basic reasons: to clear land for agriculture and settlement, and for

Francesca T. Grifo is Director of the Center for Biodiversity and Conservation at the American Museum of Natural History.

the wood itself. Wood is vital for shelter and furniture construction, and in many countries it remains the basic fuel for heating and cooking meals. The cutting of forests is a worldwide activity, from parts of Siberia and Alaska to the tropics of Asia, Africa, and the Americas. Timber harvesting and conversion to farmland have already removed almost fifty percent of the Earth's original forest cover, much of it in temperate regions. Today, these human activities account for the loss of at least sixteen million hectares of forests every year, mainly in the tropics. Such direct destruction of forests is the primary reason for today's massive extinction of species.

There are many examples of diseases which are influenced by deforestation, including malaria, leishmaniasis, Kyasanur Forest Disease in India, Chagas disease in South America, Lyme disease in the United States, and increased water-related diseases the world over.

Malaria, a potentially fatal infection, is endemic in ninety-one countries, putting about forty percent of the world's population at risk. Caused by one of the parasitic protozoans in the genus *Plasmodium*, it is transmitted to humans by a mosquito vector of the genus *Anopheles*, which thrives in moist tropical and sub-tropical habitats. In Amazonian Brazil, malaria was largely controlled in the 1960s, but epidemic outbreaks occurred in the 1980s because of massive settlement and ecological disruption in the Amazon Basin. Over half a million cases were reported in 1988. The outbreaks were partly due to the large influx of people with little or no immunity to malaria, and partly because of rainforest disturbance that allowed malarial mosquitoes which typically fed on other hosts in the jungle, to come into contact with and bite people. Additionally, road construction, runoff from land clearing, and open mining in the forests left pools of water standing, ideal conditions for unlimited breeding of malarial mosquitoes. Similar examples can be found in many parts of the world.

Leishmaniasis is actually a group of protozoan diseases transmitted by biting sandflies. The particular *Leishmania* pathogen may infect the skin, the mucous membranes, or the internal organs. Some types are fatal if untreated; others cause disfiguring scars and ulceration, and increase vulnerability to other potentially life-threatening infections. Leishmaniasis occurs in eighty-eight countries, affecting over twelve million people globally in its various forms.

Forest rodents carry the protozoan and are the usual reservoir hosts for sandflies, but when the rainforest is cleared and new villages introduced, these rodents are displaced. The flies then turn to biting humans in the absence of sufficient numbers of their preferred rodent hosts. The spread of leishmaniasis is further accelerated by road building, dam construction, mining, and other development programs that bring more people into contact with the sandflies.

Shortly after a city called Cidade Nova was built outside Manaus, Brazil, adjacent to primary forest, the incidence of leishmaniasis skyrocketed. The reason was that pathogen-bearing sandflies were able to move from the primary forest into the nearby houses and transmit the disease to human inhabitants from sloths and anteaters in the surrounding primary forest with infection rates of leishmaniasis of over fifty percent.

Kyasanur Forest disease was the first tropical tick-borne virus to be discovered. First noted in 1957, the disease peaked in 1983 with 1,555 cases and 180 deaths recorded in that year alone. It is caused by a pathogen in the genus *Flavivirus*, which is carried by *Haemaphysalis spinigera* tick vectors in the tropical forests of Mysore in southern India.

The ticks multiplied when rainforest cutting for small-scale agriculture led to the invasion of the cleared areas by *Lantana camara*, a thick brush species. Livestock were introduced to graze in the clearings and the brush, in turn, provided habitat for small mammals. Tick density increased because adult ticks could feed on the cattle and sheep while the immature ticks (nymphs), which infect humans, used the small brush-dwelling mammals as reservoirs. The abundant nymphs could then attack people coming into the clearings to gather fuel, produce or tend their livestock. First noted in 1957, the disease peaked in 1983 with 1,555 cases and 180 deaths recorded in that year alone.

Triatomine bugs (mainly Triatoma infestans) are the vectors of *Trypanosoma cruzi*, the agent of Chagas disease. They originally lived in woodland habitats in South and Central America where they transmitted the parasite among forest mammals. The bugs exploited the major environmental changes in the post-Colombian New World by becoming a widespread inhabitant of poorly sealed human houses typically built in formerly forested areas. The last one hundred years have seen dramatic changes in its distribution with resultant outbreaks of Chagas disease.

Since 1958, more than 20,000 Argentineans have been infected with Junin virus. In the 1940s, widespread clearing of central Argentine pampas (natural grasslands) to plant corn involved heavy herbicide use to control native grasses and weeds. Grasses that could live under these new conditions became wide-

A hut in Western Bolivia of the type particularly suited for harboring triatomine bugs, the vectors of Chagas disease.

spread, and replaced the native grasses. Thriving on the seeds of these newly dominant grasses, and because its natural predators disappeared when the pampas vegetation was eliminated, the local field mouse (*Calomys musculinus*), a carrier of the virus, flourished in the cornfields. As the mouse population grew way beyond their former small numbers, farmers were exposed to their feces when they reaped the corn. The fever spread, and as native grasslands were steadily replaced by monocultures, the incidence of the disease increased dramatically, affecting 40,000 square miles in 1990.

In eastern Bolivia in 1962, Bolivian hemorrhagic fever (caused by the Machupo virus) surfaced as a result of deforestation and predator removal. In the early 1950s, dense jungle patches were cleared to grow corn and other vegetables. This disrupted the natural habitat of the host of the virus, a field mouse in the genus *Calomys*. The corn provided the mice with an excellent food source. Further, all the village cats had been killed from massive DDT spraying to control malarial mosquitoes. The mouse populations flourished, overrunning settlements and spreading the virus in their urine and feces, killing up to twenty percent of the villagers. The epidemic abated when cats were reintroduced to the area.

Yellow fever persists in Western Africa—despite a vaccine—because of rainforest encroachment.

A forest remnant surrounded by lands cleared for agriculture near Azuay, Ecuador.

The disease was originally transmitted in a jungle cycle from monkey to monkey via the mosquito *Aedes africanus*. When logging was introduced, however, humans entered the cycle. People working in the forest on logging operations were bitten, and upon returning home, initiated an urban cycle whereby the disease was spread from person to person via A. *bromiliae* and A. *aegyptii*, which thrive in disturbed urban environments.

In much of the eastern United States, deer no longer have significant predators other than people, whose hunting of deer is restricted near human settlements. The result of this and other ecological changes has been a substantial increase in the deer population. Many of these deer carry the species of tick (*Ixodes dammini*) the vector of the bacteria (*Borrella burgdorferi*) that causes Lyme disease—named after Old Lyme, Connecticut, where the disease was first reported. The expanded deer population has carried Lyme disease to suburban areas and into contact with people and their dogs and cats. By 1992, Lyme disease was the most reported vector-borne disease in the United States. If the disease is untreated, it can result in arthritis of the large joints, muscle pain, and even nerve and heart damage.

THE EFFECT OF DEFORESTATION ON WATER QUALITY

The role of deforestation in disease ecology extends to water quality, and therefore to the incidence of water-related infectious disease. Deforestation has reduced natural purification and regulation of water flow and rainfall. As a result, formerly forested regions are experiencing flooding, drought, soil erosion, and agricultural runoff into clean water. Wetlands and forests also soak up and store excess water and slowly release it over time thus reducing floods and droughts. Flooding and the subsequent contamination of water supplies are causing an escalation in water-borne infectious diseases. Destruction of wetlands, watershed deforestation, and conversion to agriculture are major sources of water contamination through pollution of rivers, lakes, streams, coastal waters, and groundwater.

There are scores of water-related diseases. They include diseases caused by contamination from lack of hygiene, such as lice, scabies, and fungal infections of the skin; diseases spread by the fecal-oral route, such as cholera, typhoid, hepatitis, and polio; diseases caused by amoebas, such as giardia and cryptosporidium; diseases of water contact, such as schistosomiasis; soil-mediated infections, such as roundworms and hookworms, and tetanus; diseases that come from food-borne pathogens, such as trichinella (a nematode worm), tapeworms, and liver flukes, and food poisoning from *Campylobacter* and other bacteria; and insect-borne diseases in which the insect thrives in the presence of water, such as malaria, yellow fever, and dengue fever.

Water-related diseases are not limited to other countries. In 1993, Milwaukee experienced the largest outbreak of water-borne disease ever reported in the United States. An estimated 403,000 persons became ill with cryptosporidiosis. Runoff from dairy farms into Lake Michigan, the source of Milwaukee's water supply, is suspected as a source of the pathogen.

The EPA reported in its national water quality inventory report to congress in 1996, about forty percent of the nation's surveyed rivers, lakes, and estuaries are too polluted for basic uses, such as fishing and swimming. In 1993, community water systems supplied 26.5 million people with water that violated health-based standards. Overall, medical researchers have estimated that there are 7.1 million cases of mild to moderate water-borne illness annually in the United States and another 560,000 cases that are moderate to severe.

Without clean water supplies for drinking and hygiene, illness inevitably increases. Although it is difficult to tease apart the exact roles played by poverty and environmental degradation in the lack of a safe water supply, we know that clean water is essential to human health and that water-related diseases have not been entirely eliminated even in wealthy countries.

CONCLUSIONS

The Earth has experienced five mass extinction episodes over the last 500 million years. These mass extinctions resulted from terrestrial and extraterrestrial events. We are now in the midst of the sixth mass extinction, as species vanish in a time span of mere decades rather than millions of years.

We humans are the sole cause of this sixth mass extinction. Our population is now roughly 5.8 billion and it increases by roughly ninety million each year. Overconsumption of our natural resources compounds the environmental problems resulting from this rapid population growth. A mere one fifth

Runoff from a field cleared for cattle grazing. Less than a mile from where this photo was taken, this small stream connects with a larger stream which is the water supply for several towns in western Bolivia.

of the world's people live in developed nations such as the United States, but their per capita consumption rate is ten to one hundred times greater than that of the four fifths of the world's people who live in developing countries such as Bolivia or India.

For the first time in the history of the planet, we are altering the basic chemistry, physics, and physiology of the Earth. We are changing the atmosphere, the oceans, and the land. We are overexploiting natural resources, destroying and degrading the habitats of animals and plants, introducing alien species into new environments, polluting and contaminating the environment with toxic substances, and changing the climate of the globe.

Biodiversity conservation requires bold new thinking about the way we manage the Earth. Every policy and action must recognize that the resources of the Earth are finite and that nature's components are unique and, once lost, irreplaceable. That human health and well-being depend on the conservation of biodiversity is still largely unappreciated. This is particularly true with regard to the role biodiversity plays in maintaining the balance between the human and microbial worlds. When that balance is upset, humans are the losers. As we have already seen, the results include emerging and reemerging diseases, greater virulence in existing diseases, and increased exposure of human populations with little or no defenses against the agents of infectious disease.

How has biodiversity provided an advantage in the control of pests, pathogens and parasites in our environment?

How can the disruption of water-based areas (e.g., Everglades) lead to dramatic changes in infectious diseases?

Case Study
The Cultural Dimension of Malaria

Malaria is a scourge that afflicts an estimated 250 million people annually, resulting in one to two million deaths. Throughout history, malaria has been the single most devastating killer[1] and has shaped human biological evolution. It is one of the preeminent examples of how cultural practices determine our coexistence with infectious disease. Anthropologists have studied the disease for many years, investigating how cultural considerations impact on infectious disease.

Malaria is a serious, acute (often fatally so), and chronically relapsing infection. People infected with the parasites that cause malaria experience chills, fever (sometimes as high as 110 degrees Fahrenheit), anemia (reduction in the number of red blood cells), and enlargement of the spleen. The malaria parasites are transmitted from human to human via mosquitoes. To understand malaria, you must identify the three major organismal components: humans, *Plasmodium* parasites, and the mosquito vectors that transmit the parasite.

The growth and perseverance of malaria is the result of changing lifestyles. In prehistoric times when virgin forests were first cleared to create agricultural land, new ecological spaces opened for the mosquitoes that transmit malaria to inhabit. In essence, the establishment of the complex relationship between man, mosquito, and malaria parasite is a "sequel to the introduction of iron tools for clearing tropical forest."[2] Other cultural factors include the gathering of large numbers of people in high densities (i.e., as opposed to distanced communities of hunters and gatherers) and the change to rice cultivation which requires extensive use of water irrigation and provides breeding areas for mosquitoes. Both of these cultural changes allow for the continuous transmission of the parasite from human to human.

Another aspect that is important in understanding malaria is that cultural adaptations affect disease patterns. These social solutions only come to light after long, hard examination by cultural anthropologists. Peter J. Brown and M. C. Inhorn have summarized many of these solutions in a paper entitled "Disease, Ecology and Human Behavior."[3] Some of them include:

- In Vietnam, hill tribes' traditional house types may be cultural solutions to combating malaria. Because the mosquito that transmits the blood parasite from human to human does not fly above ten feet off the ground, these people have placed their sleeping and cooking quarters on platforms above the ten-foot flight ceiling of the mosquito.
- The pattern of "inverse transhumance" (just a fancy name for when a pastoral society moves their flocks to high elevations in the summer) has been suggested as an efficient cultural solution to malaria. People in Sardinia who showed this traditional pattern of yearly movement were thought to be less susceptible to malaria.
- In Sierra Leone, individuals who are highly susceptible to malaria infection through mosquito bites have adopted the cultural practice of sleeping in thick cotton cloth with their entire bodies covered. The thick cloth is impenetrable to the bite of mosquitoes and, further, protection by the cloth is used only at night, which is the prime feeding time of the mosquito.
- Some cultures have used specific dietary changes to decrease susceptibility to malaria. For instance, some Mediterranean cultures use the fava bean in their diet, which appears to have an antimalarial affect.
- Many cultures have developed herbal medicines to combat malaria. Excellent examples of these solutions come from all over the world. The Hausa of Nigeria have over thirty-one antimalarial plant medicines,[4] the people of Madagascar have used hundreds of plants to combat malaria,[5] the Inca people introduced to the world the use of quinine made from the bark of the cinchona tree to fight malaria, and the Chinese use the widespread "weed" *Artemisia annua* as an antimalarial agent.[6]

Although several solutions are listed above, many cultural practices can also enhance the possibility of disease transmission and increase disease prevalence. It has been pointed out that studies of cultural solutions must proceed beyond mere description of the phenomenon[7] and start to include rigorous analysis of the effects of cultural behaviors on infectious disease. In essence, some of the ideas about cultural solutions to disease are anecdotal and these ideas need to be quantified more accurately.

Below is a list of several steps that will increase our cultural understanding of infectious disease. All five of these considerations outlined have been influential in our understanding of malaria from a cultural perspective. The last one is most important because it means that we not only need to characterize individual human behavior and risk factors in cultural practices and everyday life, but we also need to understand the overall social and political effects of human behavior. In the case of malaria, the social problems are immense. Despite a worldwide effort to eradicate malaria initiated by the World Health Organization in 1955, "human hosts have been beset with political, economic and social problems that have made them recalcitrant to eradication measures"[8] (see WHO profile in Section 6). The political, social, and economic problems are essential to an understanding of culture's role in infectious disease.

Five Steps to Understanding Culture's Role in the World of Infectious Disease

Peter Brown and M. C. Inhorn have summarized how human disease and culture need to be examined by making the following five important points.[9] These points, when considered, allow cultural issues to be placed in a more concise and relevant framework in the world of infectious disease.

- Infectious disease occurs within ecological settings produced by interactions between the environment and human behavior. Thus, infectious disease is what is called "context dependent," meaning that if we don't understand the context of the disease, we will not completely understand its causes and effects.
- Cultural practices can and will alter ecologies and so can greatly influence human health.
- Human behavior and culture play a significant role in how diseases are established, maintained and spread. Infectious disease is especially prone to effects by human culture.
- Human behavior and cultural practices that are "solutions" to infectious disease will in general persist in populations because they are "selected for" in an evolutionary context.
- A complete understanding of the influence of human cultural practices on disease will only come through the melding together of two important factors. The first is an understanding of individual human behavior risk factors and the second is an understanding of the over-reaching political and social structure affected by disease.

1 F. B. Livingstone, "Malaria and Human Polymorphisms," *Ann. Rev. Genet* 5 (1971): 33–64.
2 H. Bleibtrue, *Evolutionary Anthropology* (Boston: Allyn and Bacon, 1969): 290.
3 P. J. Brown and M. C. Inhorn, "Disease, Ecology and Human Behavior," in *Medical Anthropology: Handbook of Theory and Method*, eds. T. M. Johnson and C. F. Sargent, (New York: Greenwood Press, 1990), 187–214.
4 N. L. Etkin and P. J. Ross, "Malaria, Medicine and Meals: Plant Use among the Hausa and Its Impact on Disease," *The Anthropology of Medicine: From Culture to Method*, eds. L. Romanucci-Ross, D. E. Moerman and L. R. Trancredi, eds. (New York: Praeger, 1983), 231–259; N. L. Etkin, "Antimalarial Plants Used by Hausa in Northern Nigeria," *Trop Doct* 27: 12–16.
5 P. Rosoanaivo, A. Petitjean, S. Ratsimamanga-Urveg, and A. Rakoto- Ratsimamanga, "Medicinal Plants Used to Treat Malaria in Madagascar," *J. Ethnopharmacol* 37 (1992): 117–27.
6 D. L. Klayman, "Weeding Out Malaria," *Natural History* 10 (1989): 18–26.
7 P. J. Brown and M. C. Inhorn, 1990, op. cit. 213.
8 C. S. Wood, *Human Sickness and Health: A Biocultural View* (Palo Alto, CA: Mayfield Publishing, 1979), 289.
9 P. J. Brown and M. C. Inhorn 1990, op. cit. 187–214.

Profile

René Dubos: Father of Modern Medical Ecology

Throughout the book, a series of profiles highlight the work of individuals or organizations who have made a significant impact in the fight against infectious disease.

René Dubos was a sickly eighteen-year-old when he entered the Institut National Agronomique in Paris, after a bout of rheumatic fever made him miss the entrance exam for the school he really wanted to attend. His worst grades were in microbiology, which he complained was "intensely boring," but it was chemistry he disliked most, and he swore he would never set foot in a lab for the rest of his life.

By the time he died in 1982 at the age of eighty-one, he had not only revolutionized microbiology and ushered in the age of antibiotics, but had profoundly influenced the manner in which science looks at the interrelatedness of all living things. In the years between, he changed the way tuberculosis is treated, inspired countless researchers (including several Nobel Laureates), invented a liquefying process that gave us Bosco chocolate syrup, and coined the phrase "Think globally, act locally," the watchword of the environmental movement, of which he was one of the founders.

How did René Dubos make the transition from a bookish but indifferent student to one of this century's most influential scientists? In some respects, his is the story of lucky accidents, but as Louis Pasteur said, "In the fields of observation, chance favors only the prepared mind." There is no question that Dubos was prepared to seize every opportunity that presented itself and follow his curiosity wherever it led him.

Dubos was living in Rome and writing technical papers on agronomy when he chanced upon an article that said microbes should not be studied in artificial cultures but under natural conditions where the environment and other organisms influence their activities. The idea intrigued Dubos (possibly because it offered a way out of the laboratory), and changed his mind about microbiology as a subject unworthy of study. That idea led to what became the governing principle of Dubos' work throughout his life: Whether it is disease or human progress, the answers lie in the connection between living organisms. The future of the Earth, he believed, depends on it.

Before beginning his studies in microbiology, Dubos wanted to visit America. A fellow passenger on his New York-bound ship was an American soil bacteriologist who offered young Dubos a fellowship at Rutgers University. Three years later he had his doctorate, but when he applied for a National Research Council fellowship, he was turned down because he was not an American citizen. Along with his rejection letter, however, came a note suggesting he look up Alexis Carrel, a Nobel Prize-winning physiologist at The Rockefeller Institute (now The Rockefeller University). Carrel in turn introduced him to Oswald Avery, who later discovered the role of DNA in transferring heritable characteristics in bacteria.

At the time Avery was working on type III pneumococcus, which causes a particularly virulent form of pneumonia. He was frustrated in his attempts to find a way to attack this bacterium, which was protected by an apparently impregnable polysaccharide (sugar) coating. He challenged his young visitor to devise a way to pierce the armor, and Dubos found himself with a job at one of the most prestigious research facilities in the world.

It took Dubos three years to find the answer, but what guided his search was a belief in the interaction of organisms. He knew that, in nature, organic matter is continually recycled by microbes, and the polysaccharide coating would be no exception, he was sure. His search for a microbe whose specialty was degrading polysaccharides took him to a New Jersey cranberry bog and a bacterial enzyme he called SIII. Once SIII digested the armor of type III pneumococcus, this infectious agent was vulnerable to attack by the body's own disease-fighting cells.

Dubos applied the same approach—looking for soil microbes to solve biomedical enigmas—to isolate *Bacillus brevis*, a microbe that attacks pneumococcus, staphylococcus, and streptococcus, among other common disease-causing bacteria. He extracted the antibacterial chemical in *B. brevis*, which he called tyrothricin, and identified its two components, which he named gramicidin and tyrocidine.

Gramicidin was the first antibiotic discovered in a systematic fashion, and its discovery in 1939 heralded the age of antibiotics, which has transformed modern medicine. Alexander Fleming had discovered penicillin a decade earlier, but it was quite by accident. Dubos' achievement was to develop what is referred to as a rational approach to drug discovery, and it paved the way for researchers who found all the antibiotics in use today.

But even before antibiotics became widely used to treat infections, René Dubos understood that bacterial resistance was inevitable. In 1942, he wrote "... keep in mind that susceptible bacterial species often give rise by 'training' to variants endowed with great resistance to [antibacterial] agents."[1] He also predicted that new infectious diseases—bacterial and viral—would emerge as worldwide social conditions changed the environment that humans and microbes inhabit.

It was characteristic of Dubos to move on once he had broken new ground. Perhaps he needed a new challenge, but he was content to leave the territory to others, hoping they would harvest the seeds he had planted. By the 1950s, Dubos had shifted his focus to medical ecology, and ultimately to the larger picture of the health of the Earth. In the last decades of his life, he lectured widely about environmental concerns, urging the responsible use of science and technology and sounding an early warning about humankind's impact on the planet. It was a natural outgrowth of his belief that disease must always be studied as part of an ecosystem, which was based on the seed of an idea planted by the article he read so many years before.

1 Dubos, R., "Microbiology," *Ann. Rev. of Biochemistry*, 11 (1942): 659–78.

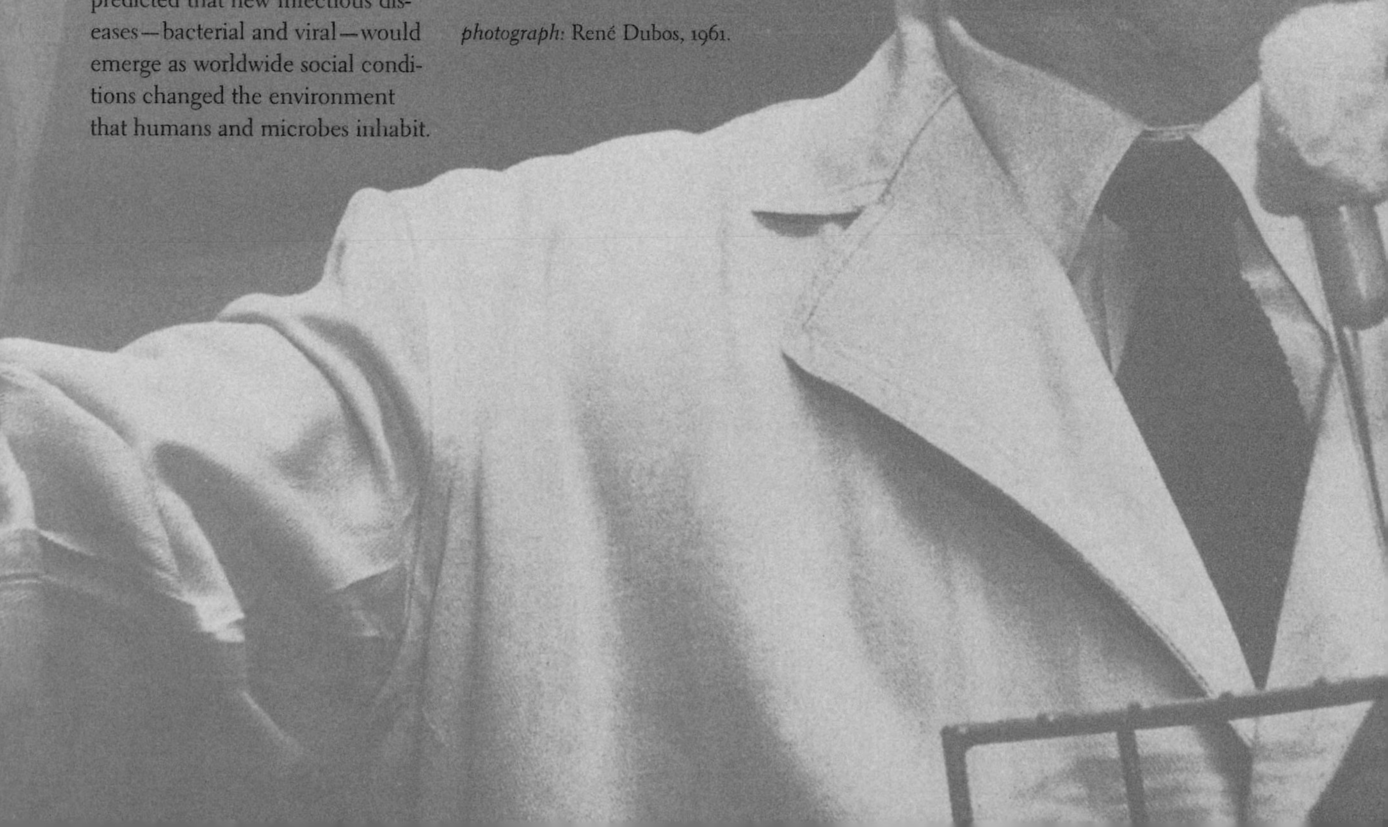

photograph: René Dubos, 1961.

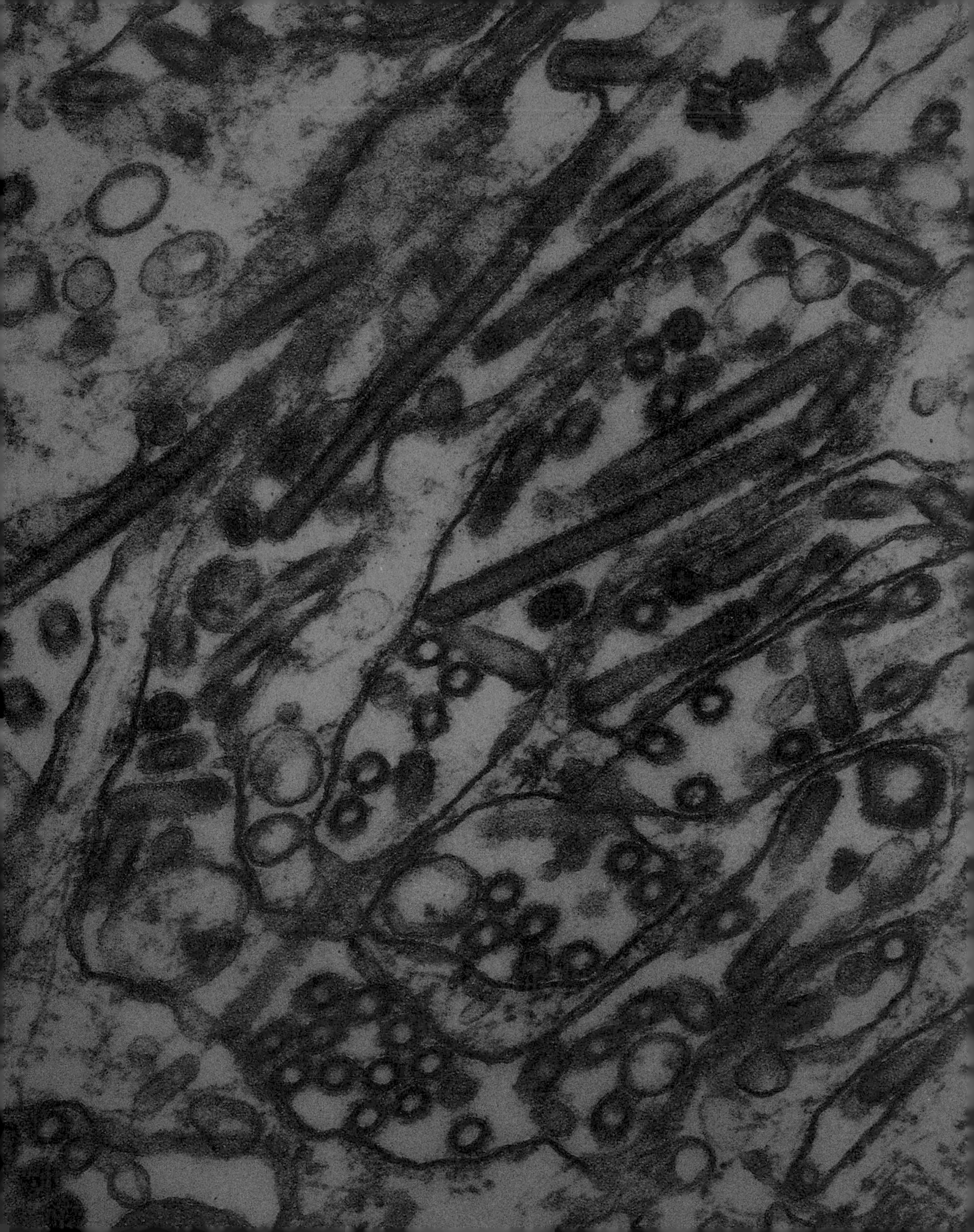

Section Two: Exposure

left Electron micrograph image of the Avian influenza virus.

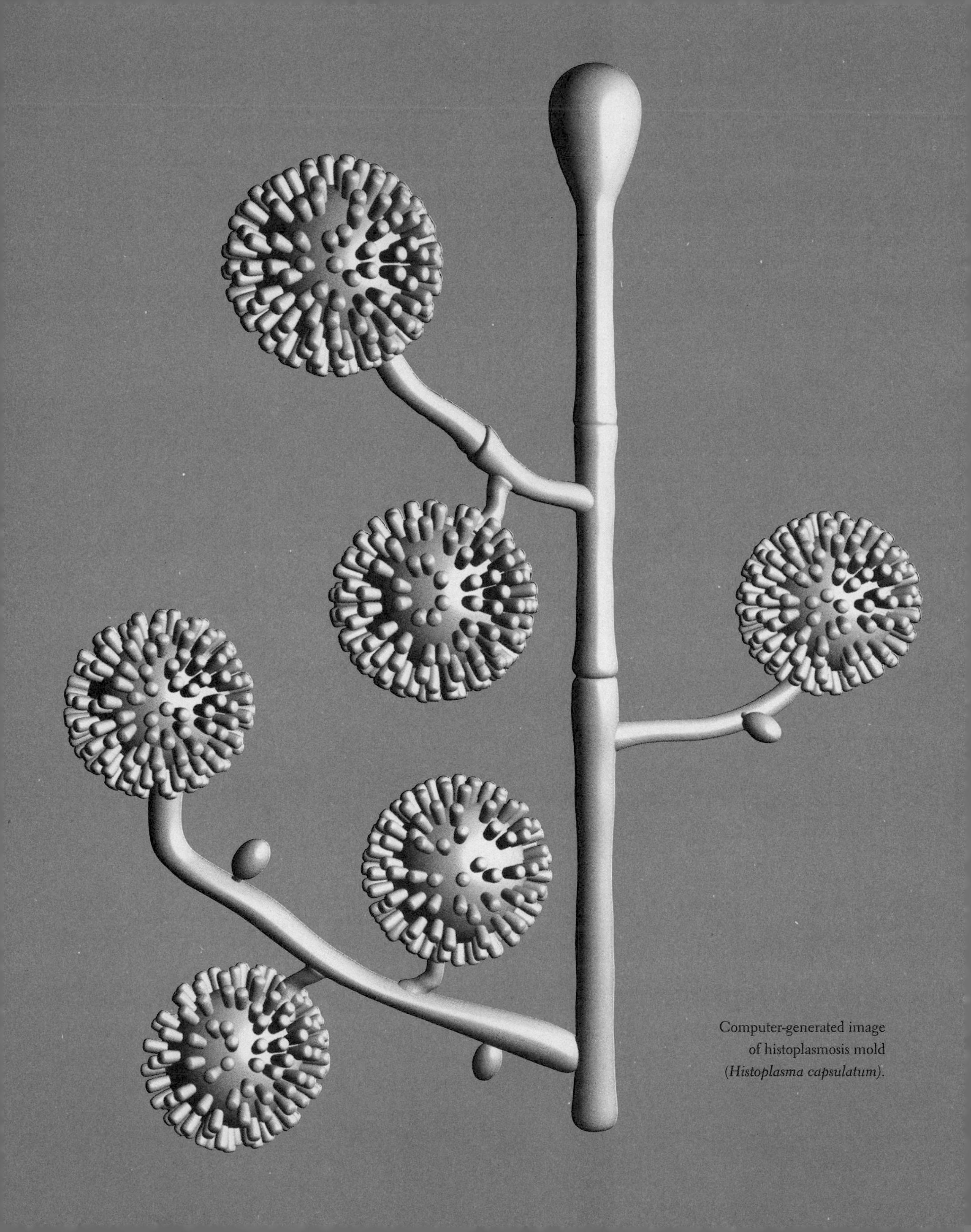

Computer-generated image
of histoplasmosis mold
(*Histoplasma capsulatum*).

Section Two: Exposure
Introduction

ROB DESALLE

In the first section, we examined infectious disease from the perspective of ecology, evolution, and culture. Now let us turn to the topic of exposure, specifically human exposure to pathogens. Using our concentric ring analogy, exposure is the first and closest ring to humans. Most medical textbooks define exposure as "When microbes meet [hu]man." To understand how, when, where, and why humans meet infectious microbes, we need to know something about the "whats"—the microbes themselves. Awe-inspiring in their diversity, microbes are not one group of organisms; many important microbes are not even living things. It is this great diversity among microbes—manifested by a diversity of structure, modes of reproduction, ecological relationships with other species, and evolutionary history—that makes the subject of human exposure to them so challenging.

Although humans and microbes have had an intimate relationship for millions of years, Anthony van Leeuwenhoek was probably the first person to see a microbe—a mere 300 years ago. By the nineteenth century, Anthony van Leeuwenhoek's hand-held microscope had evolved into a powerful tool that led to a revolution in biology. In the 1830s, Matthias Schleiden and Theodore Schwann laid the theoretical foundation of modern biology when they established that the cell was the basic unit of all living things, which, they noted, could be either unicellular or multicellular. In 1855, Rudolph Virchow developed the idea of "cellular pathology," that is, diseases are diseases of cells. In the latter third of the nineteenth century, Robert Koch and Louis Pasteur, the first modern microbiologists, recognized the ubiquity of microbes, and that they frequently inhabited the cells of other living things. They provided solid evidence that microbial lifestyles were responsible for such staples as bread and wine, and for such diseases as anthrax, cholera, and tuberculosis. When microbial lifestyles go awry, they usually wreak havoc, causing, for example, wine to turn sour, meat to putrefy, and disease to occur. Pasteur and Koch realized that these sorry consequences often followed a disruption—caused by invading microbes—of the community of resident microbes.

The more scientists learned about microbes, the more confident they became that they could control them. Pasteur developed a method to retard the souring of wine and the spoiling of milk: heat them just enough to destroy any microbial invaders and then keep them sealed. Like this "pasteurization process," the early attempts to control harmful microbes were usually mechanical, derived from an understanding of the lifestyles of these microscopic foes. These early antimicrobial efforts rapidly benefited public health; for example, the pasteurization of milk significantly reduced the incidence of tuberculosis, which had spread in part through tubercle bacilli in cows' milk.

During the first half of the twentieth century, incredible advances in microbiology came once again from scientists who attempted to understand how microbes live and reproduce. Scientists including Selman Waksman, Oswald Avery, René Dubos, Alexander Fleming, and Gerhard Domagk capitalized on their vast knowledge of the lifestyles of microbes to produce the world's first antibiotics. These natural and synthetic antibiotics were heralded as the agents that would end the war on infectious disease. During this same period, we became aware of the existence and striking characteristics of a new group of microbes much smaller than, and fundamentally different from, bacteria—the viruses.

Three major groups of microbes—bacteria, viruses, and protists—cause most infectious diseases. The incredible diversity among these microbes results in equally varied modes by which they meet humans. We need to examine the what, when, where, how, and why of exposure to our microbial enemies in order to understand infectious disease.

WHAT? Three important aspects of microbes—their structure, mode of reproduction, and evolutionary history—are crucial factors affecting both the human exposure to microbes and the procedures used to diagnose specific infectious diseases. Microbes are incredibly diverse in their structures and lifestyles, and not all of them are harmful to humans. In fact, the vast majority of bacteria, viruses, and protists either do not affect humans or are beneficial to them. The incredible structural diversity among microbes has contributed greatly to their evolutionary success. At the same time, it has also enabled physicians to identify them in infected patients and then administer medicines that destroy them without harming non-infected human cells. The modes by which microbes reproduce are also incredibly diverse. To mention just one example, spores of the cholera-causing bacteria *Vibrio cholerae* can lay dormant in plankton as they raft across an ocean and later spread a cholera pandemic when they reach the shore. Underlying both the structure and the reproductive mode of microbes are their various evolutionary histories, presented as "family trees." For example, the trypanosomal protists that cause sleeping sickness are more akin to us than to bacteria. Understanding our evolutionary relationship to various pathogenic microbes helps guide research aimed at their identification and control.

WHEN? Humans are at an increased risk of exposure to pathogens when their internal microbial ecosystem is disturbed, when their local ecosystem is altered by a natural or human-caused disaster, by the arrival of a new species, or when people travel to a different ecosystem.

WHERE? Microbes are everywhere; they play essential roles in every ecosystem, including the human body. The exposure of humans to infectious microbes commonly occurs in such "stressed" ecosystems as cities, refugee camps, hospitals, day care centers, and devastated natural regions.

HOW AND WHY? The disruption of stable ecosystems creates ecological niches for infectious microbes. Factors that change ecosystems so that they become "hotbeds" for exposure of humans to microbes are both natural and human caused. Natural phenomena that change ecosystems include such global phenomena as El Niño events and global warming, and such local phenomena as floods and droughts. Human-caused stresses on ecosystems include deforestation, introduction of agriculture into natural regions, and devastation wrought by war.

To explore exposure, I pose the following questions:

What are microbes? How many kinds are there? How do they work?

MARLA JO BRICKMAN, the Scientific Coordinator for the exhibition "Epidemic! The World of Infectious Disease" at the American Museum of Natural History, provides an overview of the characteristics and behavior of microbes, including their structures, reproductive strategies, and interactions with humans.

What are the factors of emerging infectious diseases?

STEPHEN MORSE, Director, Program in Emerging Diseases and Assistant Professor of Epidemiology at Columbia University, looks at the effects of several key factors: ecological change, agriculture, human behavior, travel and trade, technology and industry, microbial adaptation, and public health.

How do insects spread disease?

VARUNI KULASEKERA, Research Scientist, American Museum of Natural History, examines the unique characteristics of vector-borne diseases when insects are the vector.

How do doctors diagnose and treat infectious disease?

JEREMIAH BARONDESS, President of the New York Academy of Medicine, offers a doctor's perspective on how to deal with infectious disease—a combination of being a scientist and a detective.

Together, these essays not only examine the microbes and their exploitation of opportunistic conditions, but also show how an understanding of pathogenic exposure can lead to new methods of diagnosis and treatment.

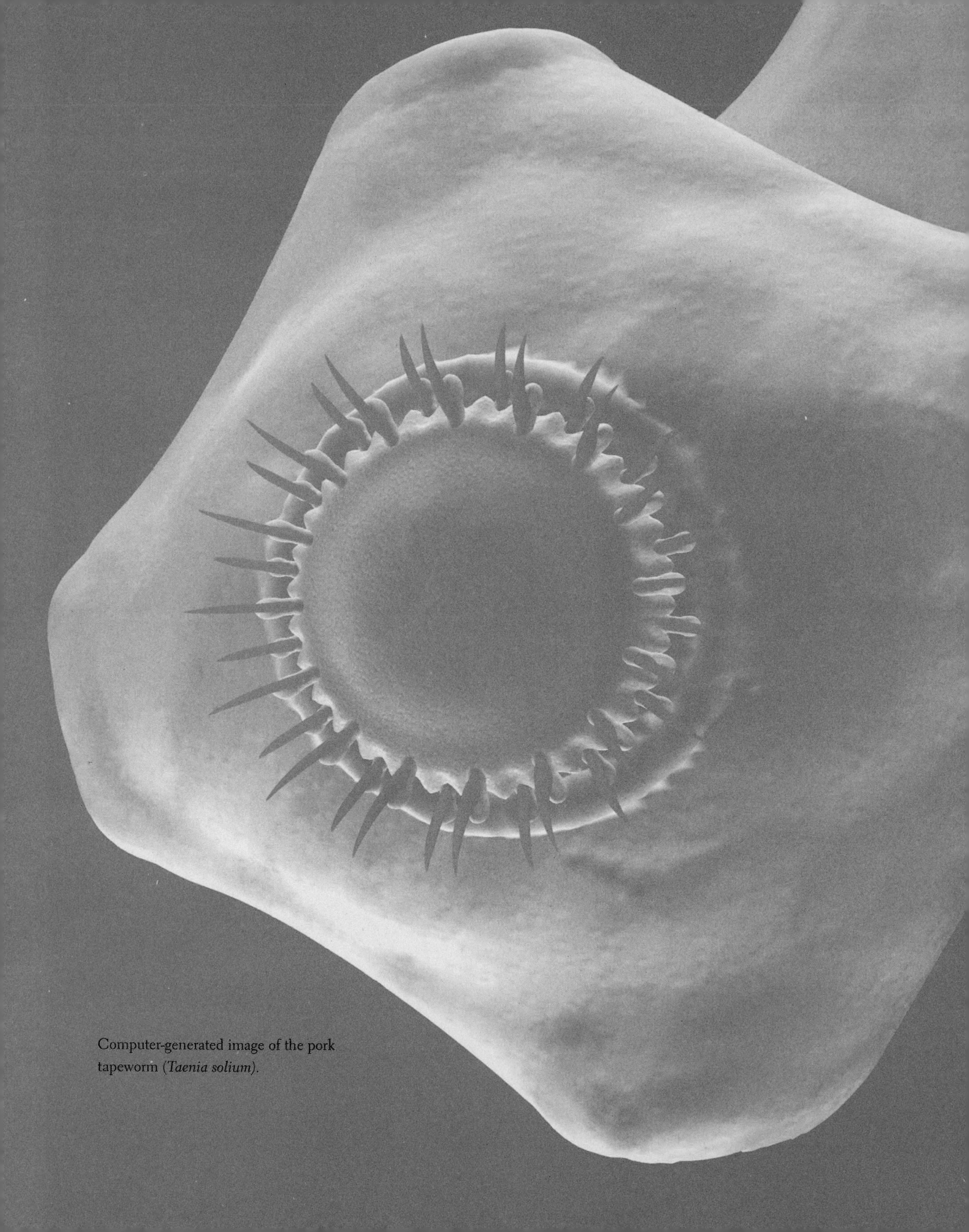

Computer-generated image of the pork tapeworm (*Taenia solium*).

Agents of Infection

MARLA JO BRICKMAN

INTRODUCTION

When we think of microbes, we think of the disease-causing "germs" or "bugs," and overlook the many ways we have exploited microbes to our benefit to produce food, make medicines, and conduct scientific research. More important, we fail to consider that the incredibly diverse world of microbes includes countless species essential to life on Earth.

The proper functioning of the human body depends on the millions of microbes that call it home. These microbes are our normal flora. They produce needed vitamins and help protect us from disease-causing microbes. We acquire our flora during birth; during the first few weeks after birth, the makeup of this flora may change through exposure to environmental microbes. The specific mixture of microbes comprising each of our floras depends on such factors as diet, living conditions, and personal habits. Once established, our normal flora remains stable—unless disrupted by disease—throughout our lives.

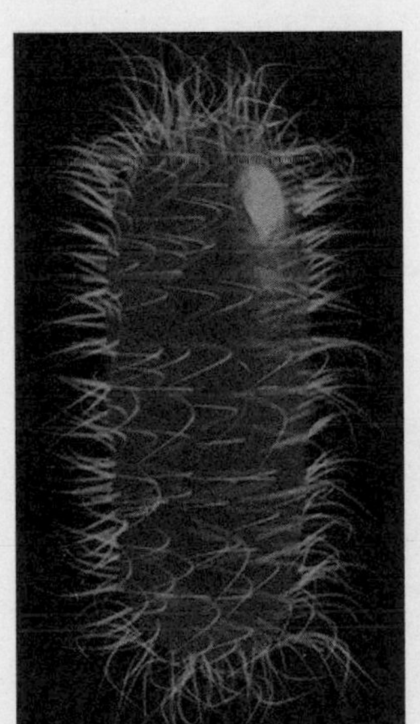

Computer-generated image of dysentery bacterium (*Shigella dysenteriae*).

Once blamed on bad air, evil spirits, angry gods, or certain groups of people, the microbial causes of infectious diseases were discovered only in the latter half of the nineteenth century, primarily through the work of Louis Pasteur and Robert Koch. In his now-famous "Koch's Postulates," Koch provided criteria to link a specific microbe to a specific disease: the microbial suspects had to be isolated from an afflicted animal, then grown on a laboratory medium, and then used to infect another animal. If the second animal became ill and identical microbes could be isolated from it, then that microbial species caused that particular illness. By identifying a disease in terms of the microbes that produced it, treatments and cures could be affected.

By the late nineteenth and early twentieth centuries, scientists had identified bacteria, fungi, and proto-

Marla Jo Brickman was the Scientific Content Coordinator for the exhibition "Epidemic! The World of Infectious Disease" at the American Museum of Natural History.

zoa as the causes of many infectious diseases. These organisms are visible using a simple light microscope and are commonly grown in the laboratory. By the turn of the century, researchers suspected that "filterable agents" (particles smaller than those observable using a light microscope and that could pass through a filter equipped to trap bacteria, fungi, and protozoa) caused certain diseases. However, not until the invention of the electron microscope in the 1930s, and the refinement of laboratory techniques, did scientists conclusively identify viruses. Today, thanks to modern techniques of molecular biology, researchers are constantly identifying previously unknown species of microbes.

Using optical light and electron microscopy, as well as modern techniques of genetic analysis, biologists have invented a useful classification system for the world's microbes. This system will be used in the following brief introduction to the world of microbial pathogens. Biologists divide microbial organisms into two main groups: the Prokaryotes and the Eukaryotes. Prokaryotes are primitive organisms such as bacteria that lack a truly defined nucleus in their cells. The cells of Eukaryotes, which include protozoa, fungi, plants, and animals, have a membrane-bound nucleus. Viruses and prions are not considered cells. Their evolutionary history, and possible relationship to organisms, is still unclear.

Computer-generated image of spores of *Histoplasma capsulatum* mold.

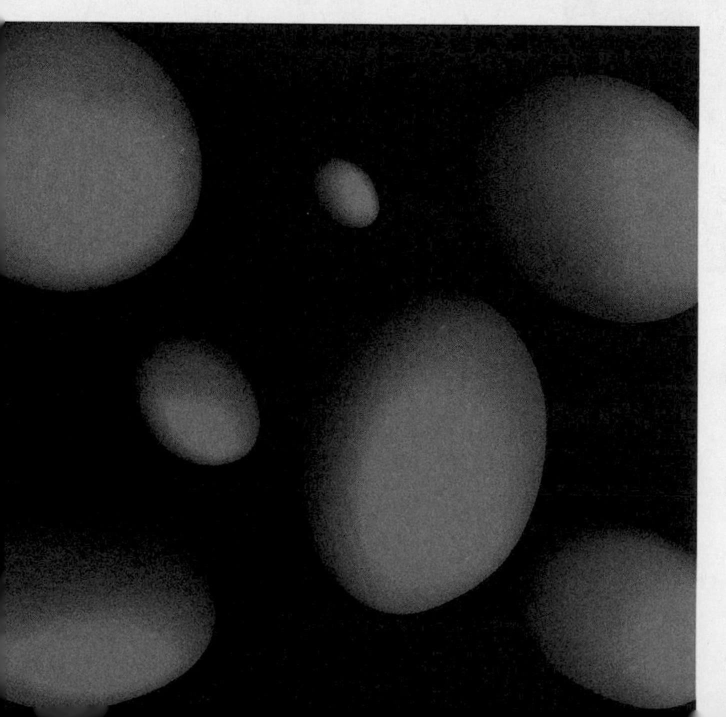

BACTERIA

Bacteria are single-celled Prokaryotes. Although each bacterium lacks specific internal structures, it nevertheless can carry on all the essential functions of life. Most of the thousands of species of bacteria are harmless—only a few are human pathogens—and some are crucial for the continuance of life.

The genetic material in bacteria is deoxyribonucleic acid (DNA), which occurs mainly as one large circular structure. Bacteria may also contain other smaller loops of DNA called plasmids. Some plasmid DNA contains genetic information that produces such traits as resistance to antibiotics or the production of a toxin.

Bacteria occur in three general shapes—spheres, rods, and spirals. Just inside the cell wall is the cell membrane through which the transport of substances into and out of the cytoplasm (the fluid part of the cell) can occur. The cytoplasm contains no organelles (subcellular "organs") other than ribosomes (tiny protein-manufacturing "factories"). It is the location where the metabolic reactions that provide the cell's energy and structural materials occurs and where the replication of its genetic material required for its reproduction occurs. Some species of bacteria have capsules that store nutrients and cellular waste. Some species of bacteria use flagella to move. Filamentous structures called pili can anchor a bacterium to a surface or can transfer genetic material between cells.

Bacteria are the most abundant organisms on Earth: one gram (1/25 ounce) of soil can contain

Computer-generated image of diarrhea-causing protozoan *(Giardia intestinalis)*.

more than a billion bacteria. They occur in all environmental niches—including soil, water, and people. Aerobic bacteria require levels of oxygen typical of the lower atmosphere. Anaerobic bacteria can survive and grow only in levels of oxygen that are much lower. Extremophiles can live in such harsh environments as hot springs or very salty seas. Other species survive as (nonmetabolizing) dormant forms called spores when the environmental conditions become too harsh. When conditions improve, the spores produce the normal bacterial forms.

Certain types of bacteria can make their own "food" from inorganic molecules and solar energy. These types of bacteria are called autotrophic. Other types of bacteria called heterotrophs absorb organic molecules, such as carbohydrates. Lipids and proteins then break these molecules down to meet their metabolic needs. Bacteria can "feed" on dead or decaying matter or on living things. These types of bacteria are called heterotrophic.

Asexual reproduction in bacteria occurs by binary fission; this is a process by which a "parent" bacterial cell divides into two identical "daughter" cells. Bacteria typically reproduce every twenty minutes. One bacterial cell can, therefore, produce more than a billion cells in ten hours. Bacteria can reproduce sexually as well. In a process called conjugation, DNA-containing plasmids from one cell are transferred to another cell through the pili.

VIRUSES

Viruses are found in all groups of living things—from bacteria and fungi to plants and animals. Much smaller than bacteria, viruses can appear as helices, icosahedra (symmetrical structures with twenty plane faces), or even more complicated forms. They consist mainly of genetic material—DNA or RNA (ribonucleic acid)—which may occur in a single or double strand, depending on the species. Viral particles are not cells, however, and cannot carry out life functions on their own.

Some viral particles are "naked," consisting only of genetic material surrounded by a protein coat called a capsid. (The structural elements comprising the capsids are capsomeres, which consist of numerous identical structural units made from proteins.) Other viral particles contain an envelope made of proteins specific to the virus plus lipids and carbohydrates from the host cell. Spikes consisting of protein combined with carbohydrate stick out and help the virus first attach to a vulnerable site (called a receptor) on the host cell and then enter it. The viral particles also contain enzymes needed for its reproduction.

Once inside the cells of other species, viruses use the host cells' organelles, DNA, and other organic molecules to produce numerous copies of themselves. (For example, a single viral particle of the virus that causes polio can produce 10,000 new viral particles within a single host cell.) Viral methods of reproduction are often complicated and can vary according to the type of virus: they typically involve a complex mode of disassembly of the entering viral particle, multiple replication of its genes, and assembly of multiple new viral particles. The host cell often dies because the virus takes over the host cell's organelles in reproducing itself, or because of damage caused when the numerous newly formed viral particles break out of the host cell.

PROTOZOA

Protozoa (also called protists) are Eukaryotes whose cells lack the cell walls characteristic of the cells of fungi and plants. Protozoa are organisms composed of a single cell that includes a nucleus. Other struc-

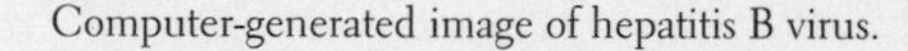
Computer-generated image of hepatitis B virus.

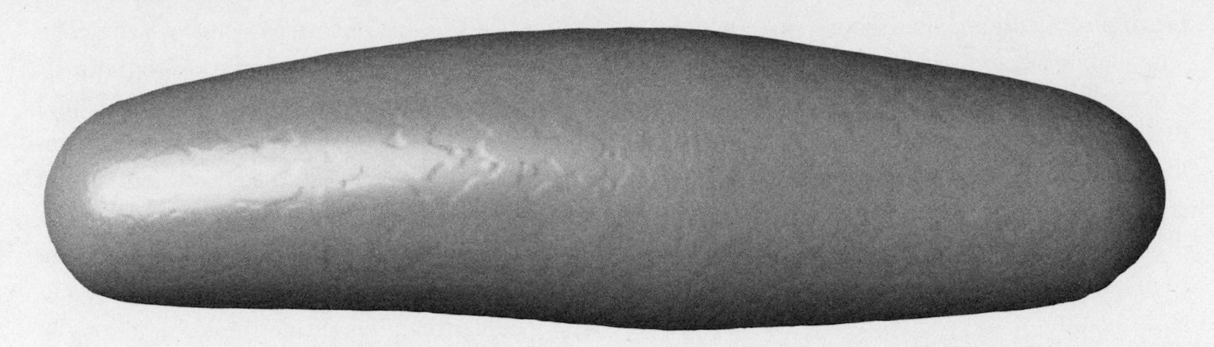

Computer-generated image of the tuberculosis mycobacterium (*Mycobacterium tuberculosis*).

tures inside the cell carry out specific processes needed for life functions. A diverse and complex group, protozoa differ in size and shape according to the species. Some, such as paramecia, are free-living in moist habitats; others, such as those that cause malaria, are parasitic, needing to live within another organism. Many protozoan species have a complex life cycle. For example, the protist may exist in one form in a mammalian host at one stage of its life cycle, and in a different form in another host, such as a fly, mosquito, or snail, at a different stage.

The components and structure of protozoan cells are very similar to those of human cells. Therefore, treating protozoan infections is difficult because drugs that may destroy the protozoan may also destroy human cells. Some basic structures found in protozoan cells are:

- a nucleus, which contains the hereditary information for the cell;
- a cell membrane, which takes in food and sends out waste;
- the cytoplasm;
- mitochondria, the energy-producing structures of the cell;
- ribosomes, structures which function in the assembly of proteins;
- the endoplasmic reticulum, an extensive array of intracellular membranes that contain ribosomes;
- Golgi apparatuses' which are involved in modifying proteins;
- lysosomes, organelles that contain enzymes that digest food and cellular debris;
- flagella, pseudopods, or cilia, which are used for locomotion and/or attachment.

Protozoa can get food by a variety of mechanisms depending on the species. Some protozoa can produce food from water and carbon dioxide through the process of photosynthesis; most feed on other organisms or organic materials in their environments (their host in the case of parasitic protozoa).

Some protozoa reproduce asexually through such processes as binary fission (to form two identical daughter cells), multiple fission (to form many offspring that may not resemble the parent), or through the formation of buds that develop outside the parent cell. Some reproduce sexually.

FUNGI

The most familiar fungi—mushrooms and molds—are visible to the naked eye. However, most species of fungi that infect healthy humans are microscopic, as are the (one-celled) yeasts we use to make bread and alcoholic beverages. The fungi are a kingdom of Eukaryotes whose cells have walls (like plant cells but unlike animal cells) but lack chlorophyll and therefore do not make their own food. Unlike typical protozoa, which engulf their food and then digest it internally, fungi digest their food externally: they secrete digestive enzymes into their immediately surrounding environment and then absorb some of the products of this enzyme-catalyzed digestion. Through this type of digestive process, many fungi play a crucial ecological role by breaking down the remains of dead plants and animals. Fungi reproduce by various methods, both sexual and asexual, including budding, fragmentation, and spore formation.

PRIONS

Still not well understood, prions (short for proteinaceous infectious particles) are abnormal forms of proteins found in mammalian brains. Prions are strikingly unlike any other infectious agent because they have no genetic material. When a prion particle infects an animal brain, it can cause the normal brain protein to change to the prion's abnormal form. The result can be progressive destruction of brain cells, resulting in a brain filled with a sponge-like pattern of holes. Among the "spongiform" diseases that affect humans are bovine spongiform encephalopathy (mad cow disease), Kuru (in which the infecting prions spread through the ritual cannibalism of the brains of the deceased), and Creutzfeldt-Jakob disease.

HELMINTHS

Not all human pathogens are microscopic, and therefore microbes. Several important and decidedly macroscopic pathogens belong to the group of animals called helminths or parasitic worms. The helminths include three main groups of animals (which are multicellular Eukaryotes): the flukes and the tapeworms, which comprise two separate classes of parasitic flatworms (phylum *platyhelminthes*), and the parasitic roundworms (phylum *nematoda*). The human "parasites" among these are so called because they live in humans, harming them in the process. Some helminths grow inside the human body to become the largest parasites known to cause disease in humans: certain species can reach a length of ten yards! Because of their size, they live outside, not inside, cells. Many have complex life cycles, living at various stages in different host species.

The body of a fluke is oval and covered by a tough outer body layer called a tegument. The tegument covers layers of circular, longitudinal, and diagonal muscles, and protects the fluke from harsh environments such as the human digestive tract. Flukes are hermaphroditic; that is, they contain both male and female sex organs. Most often, fertilization occurs mutually between two flukes, although, at times, self-fertilization can occur. Many flukes have a complex life cycle that can involve from one to several hosts.

Unlike flukes, tapeworms have no mouth, digestive tract, or even digestive enzymes: they absorb predigested food directly through their tegument. The worm has a head (called a scolex) that attaches the worm firmly to its host's digestive tract, and the body contains hundreds of segments called proglottids. Each proglottid is a sexually complete hermaphroditic unit. Cross-fertilization between two different worms occurs when possible, but when not, one proglottid can fertilize another on the same worm.

Roundworms have two distinct sexes; males are smaller than females. These worms reproduce sexually. Roundworms can be found in all types of environments, from polar regions to tropical regions. Although some are parasitic—on plants or animals—many free-living species occur in aquatic environments, including fresh water, sea water, and water in the soil.

SUMMARY

Microbes are quite diverse; this diversity is large even among members of the same microbial species. Diversity greatly increases the odds that species of microbe, plant, animal, and people will survive because it enables the species to adapt to changes in its environment. Because of the short time microbes require to mature and have offspring, and because they often can have many offspring, microbial populations adapt very quickly to change. This is why many species of microbes are now resistant to drugs that once killed them even though the drugs have been around less than fifty years.

Because humans generally regard microbes as evil predators, we have spent enormous amounts of energy and resources to try to rid ourselves and our planet of these organisms. Like us humans, microbes are species that share the drive to survive long enough to reproduce. It is true that our elimination of certain microbes has relieved human misery without, as far as we can tell, harming the Earth. However, there are thousands of species of microbes—some known, most unknown—whose survival might be vital to the survival of the global ecosystem. We need to learn to live with these unseen organisms, because without them, there probably would be no life at all.

How has the acceptance of "Koch's Postulates," criteria to link a specific microbe to a specific disease, changed how we approach the study of infectious disease?

What do the major groups of microbes have in common and what makes them distinct?

Computer-generated image of a flatworm (*Planaria dugesia*).

One of the most remarkable stories about infectious disease features a little-understood pathogen, two Nobel Prizes (one of which some people think should not have been awarded), mad cow disease, and a cannibalistic funeral rite.

It all began in the mid-1950s, when D. Carleton Gajdusek, a researcher for the National Institutes of Health, became intrigued by a peculiar disease seen among certain tribespeople in the Fore highlands of Papua New Guinea.

The Fore people called it the laughing death or *kuru*, which means trembling, an apt description of its earliest symptoms. At first the victim is beset by uncontrollable trembling, followed by loss of balance, then total loss of control of the limbs. Within a year, paralysis and dementia develop, progressively worsen, and eventually lead to death.

The Fore people believed the disease was the result of sorcery, but it has now been established that it is caused by a mysterious infectious agent called a prion. Pronounced pree-on, which stands for proteinaceous infectious particle, it is responsible for a completely new disease-causing mechanism, or so proponents of the prion theory believe. Some scientists dispute the very existence of prions, insisting that kuru and a handful of other degenerative brain diseases are caused by an as-yet unidentified "slow virus." These scientists were angered when Stanley B. Prusiner was awarded a 1997 Nobel Prize for his pioneering work on prions, because they believed that it would discourage investigation of alternative theories.

Prusiner's theory holds that the arrangement of amino acid molecules in a normal protein that he calls PrP (prion protein) and that can be found in humans and other mammals, sometimes "flips," producing an abnormal form that causes characteristic damage to brain tissue. How and why the flip occurs is not yet well understood, but researchers continue to try to answer this and other questions about prions. This accident of nature is particularly nefarious because each abnormal protein is capable of causing normal proteins to change form as well. If it sounds like a vampire scenario, read on: prions contain no genetic material and therefore are unable to reproduce. Yet they can populate the brain with their own kind through the exponential process of shape changing. Each new abnormal PrP can transform any number of normal ones, which in turn transform others, and on it goes until the host dies.

Or does it end there? In the case of kuru and its cousin, mad cow disease, that is only the beginning.

When Carleton Gajdusek set out to investigate the mysterious trembling disease, he approached the problem systematically, collecting data on the incidence of the disease, studying the habits and customs of the people in whom it occurred, and collecting tissue samples from the brains of victims. He focused on the brain because the symptoms suggested neurological damage, which he assumed was caused by a virus. But when he examined the tissues under a microscope, he saw no evidence of inflammation, which would have been present if viral infection were responsible. Instead he saw holes where brain cells should have been, and those holes were filled with protein deposits. The sponge-like appearance of the infected brains resembled those of victims of a class of illnesses known as transmissible spongiform encephalopathy (TSE).

Some TSEs affect humans: Creutzfeldt-Jakob and Gerstmann-Stäussler-Scheinker diseases affect adults, Alpers syndrome strikes infants, and fatal familial insomnia is inherited, as its name suggests. Although TSEs are invariably fatal, they are extremely rare in humans. More often, TSEs affect animals ranging from cats, mink, elk, and mule deer, to sheep and cows. The sheep disease is called scrapie, named because it causes itching so intense that sheep scrape off their wool by rubbing against any available surface. In cows, we know it as bovine spongiform encephalopathy (BSE), or more famously, mad cow disease.

The odd thing about these diseases is that they seem to be infectious (or at least transmissible) as well as inheritable, and in some cases, they appear spontaneously, without being passed by genes or by contact with an infectious agent. It is believed that a genetic mutation is responsible for scrapie; British cows became infected by eating feed that contained ground-up sheep parts, some of it from scrapie-afflicted sheep. But as Gajdusek was soon to discover, the way the Fore people were acquiring kuru was stranger still.

He noted that most of the victims were women and children, so he looked for daily tasks that were "women's work" among the Fore. He hit the jackpot when he learned that women, often helped by young children, were responsible for preparing the Fore dead for burial. Part of the burial ritual involved removing the brains of the dead, grinding them up, and cooking them as a soup, which was consumed by grieving family members. This specific form of cannibalism was practiced by other tribes in the region, but they exhibited no signs of kuru. The assumption is that at some point in the past a single Fore tribesperson acquired kuru, died of it, and his or her brain soup started the process of infection, which was perpetuated each time another victim died.

Where did that first case of kuru come from? Forty years after Gajdusek published his paper on kuru and nearly twenty years after he won a Nobel Prize for his work among the Fore people, Stanley Prusiner was awarded the Nobel for answering that question and many others. If Prusiner is right about the nature of prions, it happened by accident when a single protein, one of countless number in the human brain, "flipped." Given the odds of some sort of abnormality occurring on the molecular level over the course of an individual's life, such a scenario is hardly unlikely.

For the Fore people, the kuru epidemic ended when consumption of brains was halted in obedience of a government ban. In Great Britain, it is hoped that massive slaughter of potentially infected cows and changes in the formulation of cow feed have stemmed the spread of mad cow disease.

Throughout Europe, imports of British beef have been banned, and in the United States, the Centers for Disease Control and Prevention are maintaining surveillance, though the agency insists that there has been no evidence of BSE in this country. Right now, there is no cure for prion diseases. In fact, some people doubt that prions even exist.

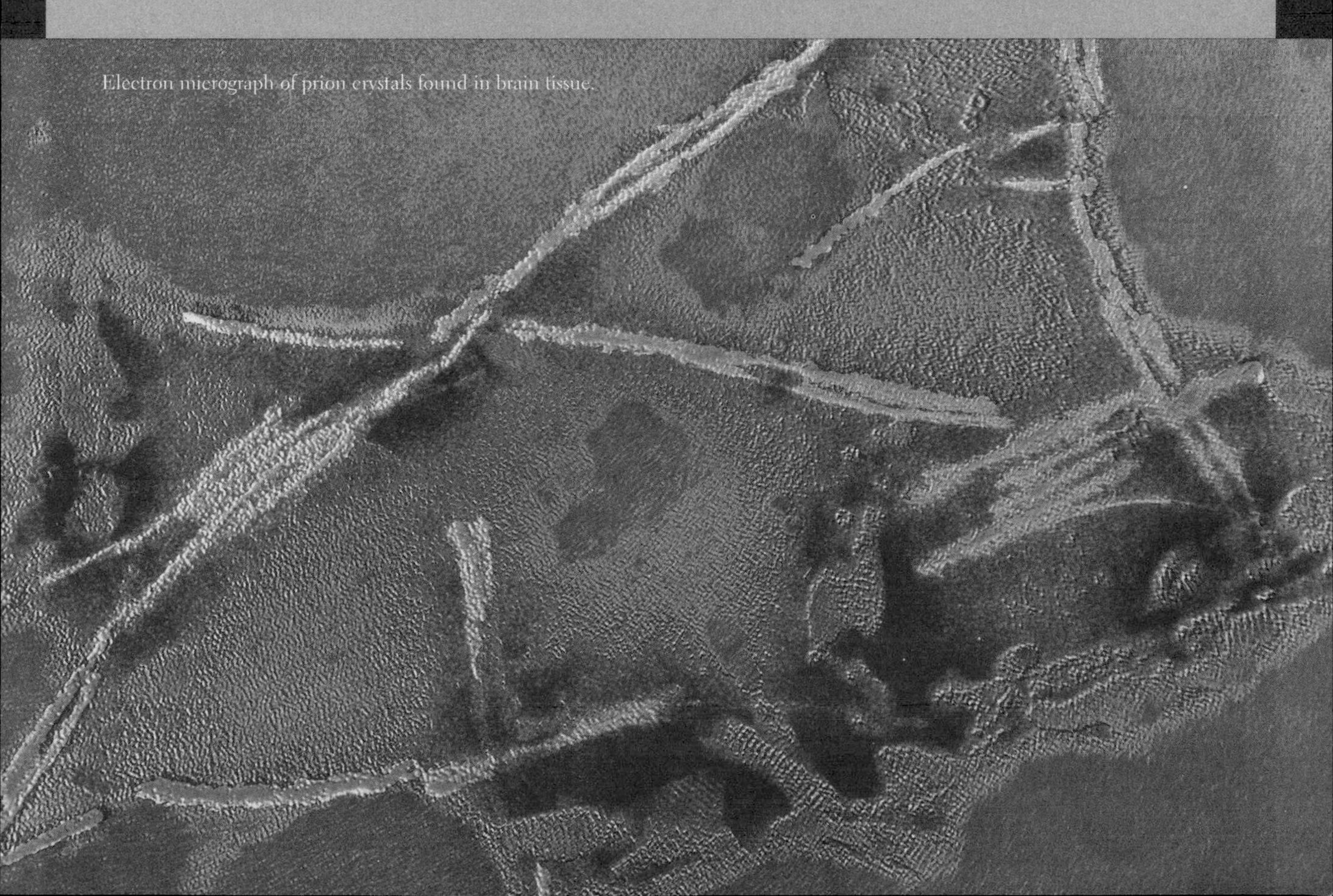

Electron micrograph of prion crystals found in brain tissue.

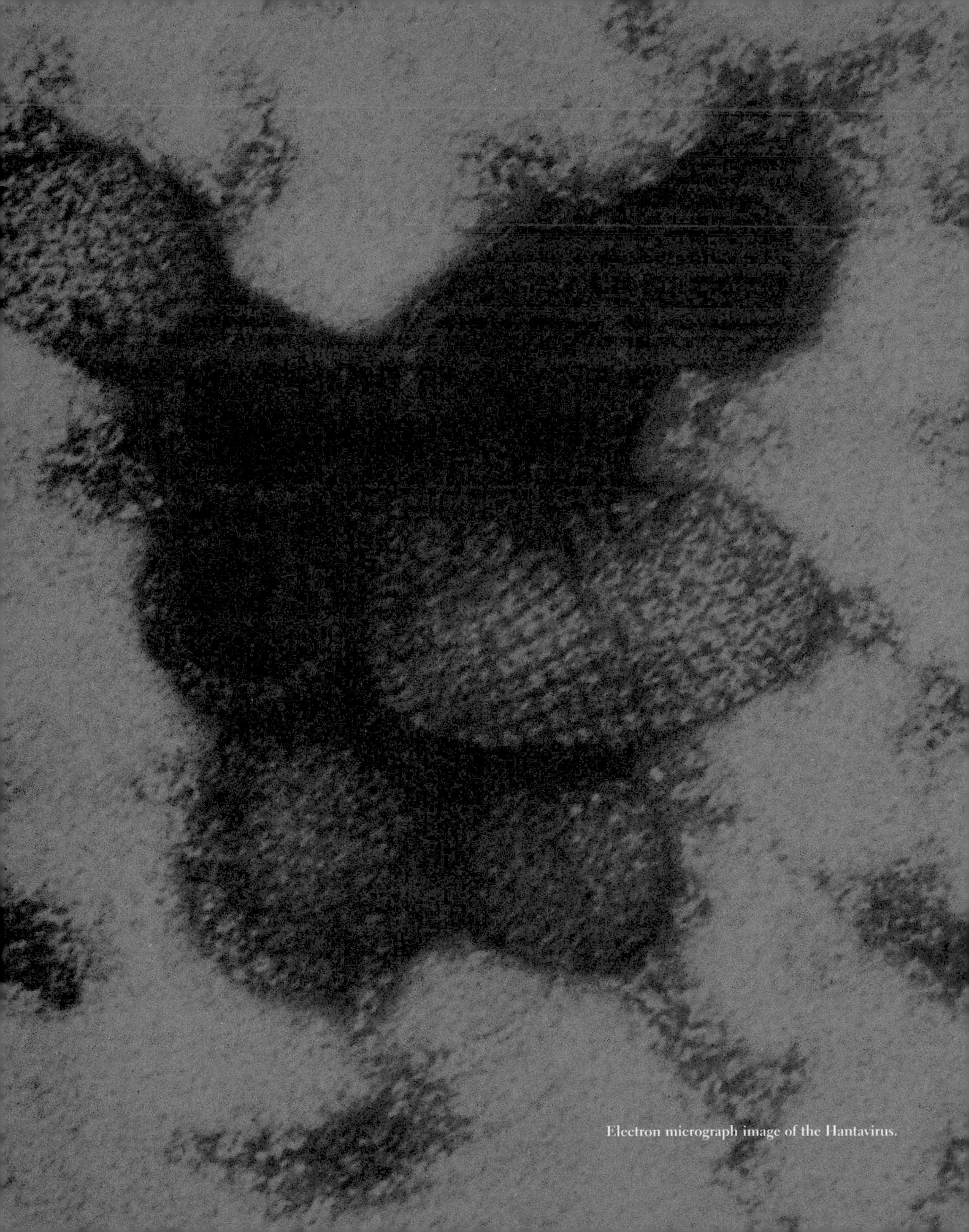

Electron micrograph image of the Hantavirus.

Factors in the Emergence of Infectious Diseases

STEPHEN S. MORSE

Emerging infectious diseases have included some of the most feared plagues of the past. New infections continue to emerge, while many old plagues are with us still. As demonstrated by influenza epidemics, a new infection first appearing at a single site may spread across the globe within days or weeks.

"Emerging infections" are those that have recently appeared, or that are rapidly increasing in incidence or geographic range. Recent examples of emerging diseases worldwide include HIV/AIDS; classic cholera in South America and Africa; cholera due to *Vibrio cholerae* O139; Rift Valley fever; hantavirus pulmonary syndrome; Lyme disease; and hemolytic uremic syndrome, a food-borne disease caused by certain strains of *E. coli*.

Table 1 (page 58) summarizes the known causes for several infections that have emerged recently. This emergence is a two-step process: (1) introduction of the infecting agent into a new host population, followed by (2) establishment and further spreading within the new host population. Factors that promote one or both steps will, therefore, promote disease emergence. Most emerging infections originate in one geographic location and then spread to new places.

Many agents now infecting humans originated in other animal species. Once a disease has appeared in humans, a variety of factors might cause it to spread from person to person. Moreover, if the species that "carries" the agent expands its range, disease caused by the agent can appear in new places. Examples include bubonic plague, transmitted by rodent fleas, and hantavirus infections, transmitted by mice.

Changing conditions may give pathogens an advantage over some competing microbe or give them an opportunity to infect new host populations. Occasionally, a new variant may evolve and cause a new disease. Both human actions and natural causes, such as climate change, can accelerate the "traffic" of microbes to new host populations.

Table 2 (page 59) summarizes the underlying factors responsible for emergence. Some of these will be considered in this essay.

Stephen S. Morse is the Director of the Program in Emerging Diseases and Assistant Professor of Epidemiology, in the Joseph L. Mailman School of Public Health of Columbia University, New York City.

ECOLOGICAL CHANGES AND AGRICULTURAL DEVELOPMENT

Ecological changes often cause disease emergence. They can place people closer to a natural host of a previously unfamiliar infection, or they can favor an increased population of the microbe or its natural host. For example, the emergence of Lyme disease in the United States and Europe was probably largely due to reforestation, which increased the population of deer and, hence, the deer tick, the vector of Lyme disease.

Agricultural development is often a factor. Hantavirus, the cause of Korean hemorrhagic fever, has occurred in Asia for centuries and currently causes more than 100,000 infections a year in China. The virus is an infection of the field mouse *Apodemus agrarius*, which flourishes in rice fields. People usually contract the disease during the rice harvest through contact with infected rodents.

Perhaps most surprisingly, pandemic influenza appears to have an agricultural origin. In China, farmers raise pigs and ducks together. Waterfowl are a major reservoir of influenza, and pigs can serve as mixing vessels for new mammalian influenza strains. Integrated pig-duck agriculture puts these two species in contact and provides a natural laboratory for making new influenza recombinants. Strains causing the annual or biennial epidemics generally result from mutation, but pandemic influenza viruses do not generally arise by this process. Instead, two influenza strains exchange gene segments to produce a new virus that can infect humans.

Infections transmitted by mosquitoes or other arthropods often result from an increase of standing water. There are many cases of diseases transmitted by water-breeding vectors; most of these cases involve dams, water for irrigation, or stored drinking water in cities. For example, the incidence of Japanese encephalitis is closely associated with the flooding of fields to grow rice. This mosquito-borne disease accounts for almost 30,000 cases, and approximately 7,000 deaths, annually in Asia. Outbreaks of Rift Valley fever in some parts of Africa have been associated with dam building and with periods of heavy rainfall.

Natural environmental disturbances, such as "regularly cyclical" weather, can also lead to disease emergence. For example, an El Niño event may have helped cause the 1991 outbreak of cholera in Latin America by encouraging a population boom of organisms that are natural reservoirs for the bacteria that cause cholera.

CHANGES IN HUMAN DEMOGRAPHICS AND BEHAVIOR

Human population movements, caused by voluntary migration or the upheaval of war, are often important factors in disease emergence. In many countries, economic conditions encourage the mass movement of workers from rural areas to cities. In this way, infections that have been isolated in small rural areas reach larger urban populations. Advances in transportation technology facilitate the spread of infectious disease from a small village to the whole world. Once in a city, the newly introduced infection spreads among the local population and then spreads farther along highways, railroads, and by airplane. HIV has been the best known beneficiary of this dynamic, but many other diseases, such as dengue, stand to benefit. Dengue hemorrhagic fever is now common in some cities in Asia. There the high prevalence of infection stems from the increasing number of open containers needed to store water for a population grown too large for its water delivery infrastructure. These containers provide breeding grounds for the mosquito vector of dengue. Too many people plus too many mosquitoes add up to the perfect environment for a dengue epidemic. Infections enjoy crowded places; hosts are readily available, transmission is assumed, and the microbes thrive. This is true not only in rural, developing countries, but also in industrialized countries where infections such as tuberculosis can spread through sites crowded with people, such as day care centers and prisons.

Human behavior can have important effects on the spreading of disease. For example, the ways in which sex or intravenous drug use has contributed to the emergence of HIV are now well known. Motivating appropriate individual behavior and constructive action, both locally and on a larger scale, is essential for controlling emerging infections. Ironically, as AIDS prevention efforts have shown, our knowledge of human behavior remains one of the weakest links in our scientific knowledge.

INTERNATIONAL TRAVEL AND COMMERCE

In the past, travel, commerce, or war might spread a geographically isolated infection to new places. For example, trade between Asia and Europe brought rats carrying the bubonic plague to medieval Europe. In recent years, opportunities for spreading infectious disease have increased because of the increasing volume, scope, and speed of traffic in an increasingly mobile world. *Aedes albopictus*, the Asian tiger mosquito, reached the United States, Brazil, and Africa in used tires shipped from Asia. Since its introduction in 1982, this mosquito has established itself in at least eighteen states of the United States. It now also carries local viruses, including the virus causing eastern equine encephalomyelitis, which is a serious disease.

TECHNOLOGY AND INDUSTRY

Modern production methods yield increased efficiency and reduced costs in operations, such as food production that processes or uses products of biological origin. However, these operations can also simultaneously increase the chances of accidental contamination and amplify the effects of such contamination. Globalization further compounds the problem. It increases the likelihood of markets to introduce pathogenic agents from far away. For example, a pathogen present in some raw material may find its way into a large batch of final products as happened with the contamination of hamburger meat by *E. coli* strains causing hemolytic uremic syndrome.

The concentrating effects that occur with blood and tissue products have inadvertently spread infections unrecognized at the time, such as HIV and hepatitis B and C. Medical settings are also at the front line of exposure to new diseases. Many infections, including many emerging infections, have spread in healthcare settings. For example, in the outbreaks of Ebola fever in Africa, many cases were hospital acquired—some through contaminated hypodermic apparatus, some by direct contact with infected patients.

Advances in diagnostic technology can lead to new recognition of agents that are already widespread. When health officials first recognize such agents, they may label them incorrectly as emerging infections. Medical science identified human herpes virus 6 (HHV-6) only a few years ago. Nevertheless, the virus is extremely widespread. Research has recently implicated it as the cause of roseola, a very common childhood disease. Because physicians have been aware of roseola since at least 1910, HHV-6 has probably been common for decades or more. Another recent example is the bacterium *Helicobacter pylori*, a probable cause of gastric ulcers and some cancers. Now physicians have successfully treated many previously intractable cases of ulcers using antibiotics.

MICROBIAL ADAPTATION AND CHANGE

Microbes, like all other biological species, are constantly evolving. Unfortunately, the widespread environmental presence of antimicrobial drugs has favored the evolutionary emergence of drug-resistant pathogens. Pathogens can also get genes that provide them antibiotic resistance from other, often nonpathogenic species. The selection pressure of antibiotics perhaps also drives this process.

Many viruses show a high mutation rate and can rapidly evolve to yield new variants. A classic example is influenza. Regular annual influenza epidemics result from minor changes in antigens in a previously circulating influenza strain. A change in a surface protein, usually the hemagglutinin (H) protein, allows the new variant to reinfect previously infected persons and their immune systems do not recognize the altered antigen.

BREAKDOWN OF PUBLIC HEALTH MEASURES AND DEFICIENCIES IN PUBLIC HEALTH INFRASTRUCTURE

Reemerging diseases are often well-recognized public health threats for which previously active public health measures such as immunization, vector control, and water purification had lapsed. Unfortunately, this situation occurs all too often in both developing countries and the inner cities of

Global Microbial Threats in the 1990s

the industrialized world. The appearance of reemerging diseases is, therefore, often a sign of the breakdown of public health measures and should be a warning against complacency in the war against infectious diseases. Cholera, for example, has recently been raging in South America for the first time in this century. Government officials may have abetted cholera's rapid spread in South America by reducing the chlorine levels used to treat water supplies. The success of cholera is often due to the lack of a reliable water supply.

FOR OUR FUTURE

The history of infectious diseases has been a history of microbes on the march, often in our wake, and of microbes that have taken advantage of the rich opportunities offered them to thrive, prosper, and spread. The historical processes that have caused the emergence of new infections throughout history continue today with unabated force. In fact, the conditions of modern life ensure that the factors responsible for disease emergence are more prevalent than ever before. The contributing effects of travel speed and global reach to disease emergence appear clearly in studies modeling the spread of influenza epidemics and HIV.

Knowledge of the factors underlying disease emergence can help us direct resources to areas in crisis and develop more effective prevention strategies. Currently, world surveillance capabilities are critically deficient. We must couple such surveillance with incentives stimulating global capabilities to respond quickly to emerging diseases. Research, both basic and applied, will also be vital in our fight against emerging diseases.

Essay adapted from Stephen Morse, "Factors in the Emergence of Infectious Diseases," *Emerging Infectious Diseases*, January-March 1995.

How has the pattern of emerging infectious diseases been influenced by humans and by changes in the environment?

What ecological and agricultural changes in developing societies might result in disease outbreaks which we could anticipate and possibly prevent?

Table 1. Recent examples of emerging infections and probable factors in their emergence.

INFECTION OR AGENT	FACTOR(S) CONTRIBUTING TO EMERGENCE
VIRAL	
Argentine, Bolivian hemorrhagic fever	Changes in agriculture favoring rodent host
Bovine spongiform encephalopathy (cattle)	Changes in rendering processes
Dengue, dengue hemorrhagic fever	Transportation, travel, and migration; urbanization
Ebola, Marburg	Unknown (in Europe and the United States, importation of monkeys
Hantaviruses	Ecological or environmental changes increasing contact with rodent hosts
Hepatitus B, C	Transfusions organ transplants, contaminated hypodermic apparatus, sexual transmission, vertical spread from infected mother and child
HIV	Migration to cities and travel; after introduction, sexual transmission, vertical spread from infected mother to child, contaminated hypodermic apparatus (including during intravenous drug use)
HTLV	Contaminated hypodermic apparatus, other
Influenza (pandemic)	Possibly pig-duck agriculture, facilitating reassortment of avian and mammalian influenza viruses Reappearances of influenza are due to two distinct mechanisms: Annual or biennial epidemics involving new variants due to antigenic drift (point mutations, primarily in the gene for the surface protein, hemagglutinin) and pandemic strains, arising from antigenic shift (genetic reassortment, generally between avian and mammalian influenza strains).
Lassa fever	Urbanization favoring rodent host, increasing exposure (usually in homes)
Rift Valley fever	Dam building, agriculture, irrigation; possibly change in virulence or pathogenicity of virus
Yellow fever (in new areas)	Conditions favoring mosquito vector
BACTERIAL	
Brazilian purpuric fever (*Haemophilus influenzae, biotype aegyptius*)	Probably new strain
Cholera	In recent epidemic in South America, probably introduced from Asia by ship, with spread facilitated by reduced water chlorination; a new strain (type O139) from Asia recently disseminated by travel (similarly to past introductions of classic cholera)
Helicobacter pylori	Probably long widespread, now recognized (associated with gastric ulcers, possibly other gastrointestinal disease)
Hemolytic uremic syndrome (*Escherichia coli* O157:H7)	Mass food-processing technology allowing contamination of meat
Legionalla (Legionaires disease)	Cooling and plumbing systems (organism grows in biofilms that form on water storage tanks and in stagnant plumbing)
Lyme borreliosis (*Borrelia burgdorferi*)	Reforestation around homes and other conditions favoring tick vector and deer (a secondary reservoir host)
Streptococcus, group A (invasive; necrotizing)	Uncertain
Toxic shock syndrome (*Staphylococcus aureus*)	Ultra-absorbency tampons
PARASITIC	
Cryptosporidium , other waterborne pathogens	Contaminated surface water, faulty water purification
Malaria (in "new" areas)	Travel or migration
Schistosomiasis	Dam building

Table 2. Factors in Infectious Disease Emergence

* Categories of factors (column 1) adapted from *ref.* 12, examples of specific factors (column 2) adapted from *ref.* 13. Categories are not mutually exclusive; several factors may contribute to emergence of a disease (see Table 1 for additional information).

Factor	Examples of specific factors	Examples of diseases
Ecological changes (including those due to economic development and land use)	Agriculture; dams, changes, in water ecosystems; deforestation/reforestation; flood/drought; famine; climate change	Schistosomiasis (dams); Rift Valley fever (dams, irrigation); Argentine hemorrhagic fever (agriculture); Hantaan (Korean hemorrhagic fever) (agriculture); hantavirus pulmonary syndrome, southwestern United States, 1993 (weather anomalies)
Human demographics, behavior	Societal events: Population growth and migration (movement from rural areas to cities); war or civil conflict; urban decay; sexual behavior; intravenous drug use; use of high-density facilities	Introduction of HIV; spread of dengue; spread of HIV and other sexually transmitted diseases
International travel commerce	Worldwide movement of goods and people; air travel	"Airport" malaria; dissemination of mosquito vectors; rat-borne hantaviruses; introduction of cholera into South America; dissemination of *V. cholerae* 0139
Technology and industry	Globalization of food supplies; changes in food processing and packaging;organ or tissue transplantation; drugs causing immunosuppression; widespread use of antibiotics	Hemolytic uremic syndrome (*E. coli* contamination of hamburger meat).; bovine spongiform encephalopathy; transfusion-associated hepatitus (hepatitus B,C); opportunistic infections in immunosuppressed patients; Creautzfeldt-Jakob disease from contaminated batches of human growth hormone (medical technology)
Microbial adaptation and change	Microbial evolution, response to selection in environment	Antibiotic-resistant bacteria, "antigenic drift" in influenza virus
Breakdown in public health measures	Curtailment or reduction in prevention programs; inadequate sanitation and vector-control measures	Resurgence of tuberculosis in the United States; cholera in refugee camps in Africa; resurgence of diptheria in the former Soviet Union

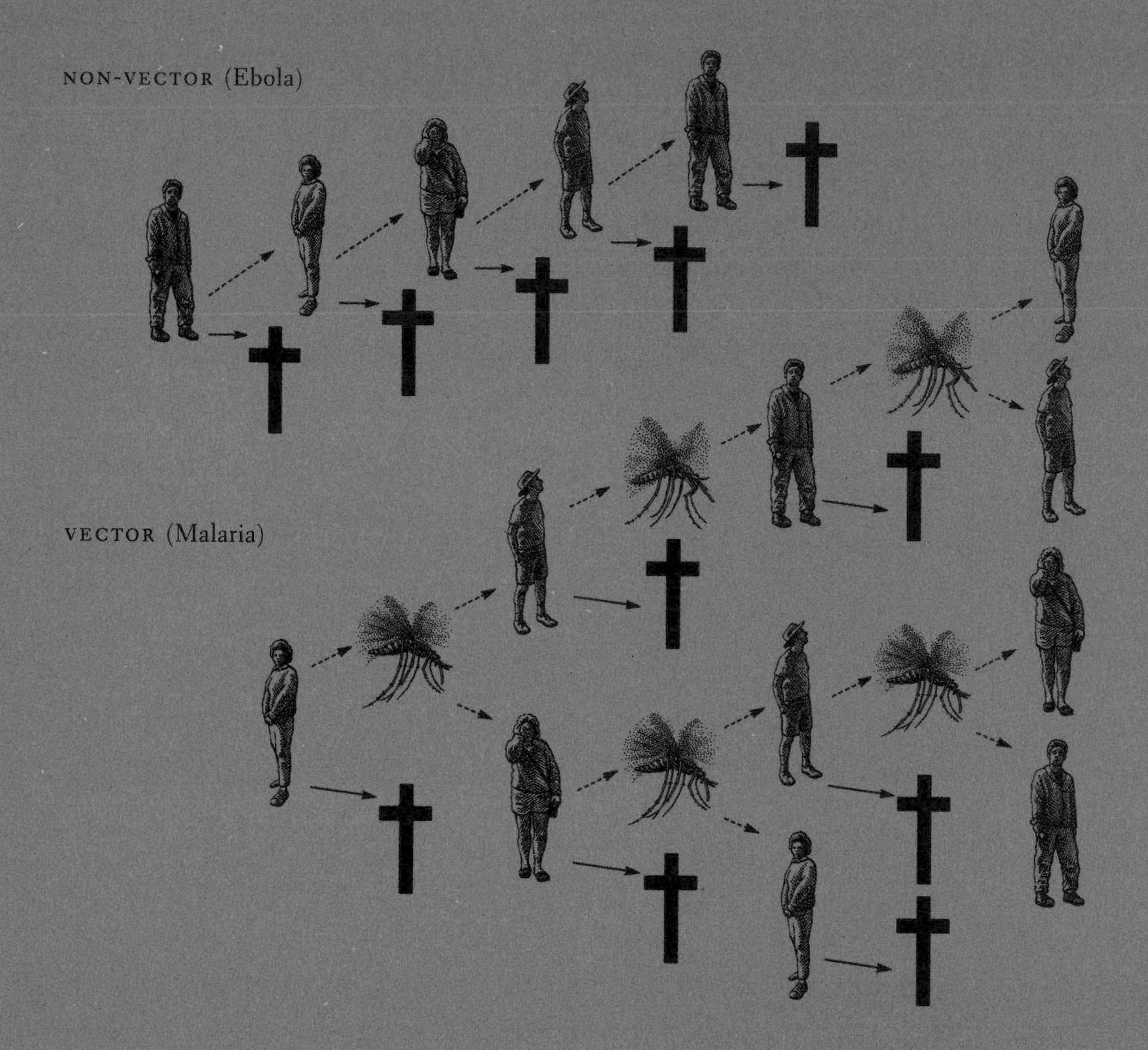

Microbes, like all life forms, want to ensure their survival. Though ebola and malaria are both highly virulent diseases, malaria has evolved a more effective strategy for increasing its chances for survival.

Ebola is a nonvector-borne disease that is transmitted through the blood. The microbe that causes ebola (a virus) kills its victims so rapidly and effectively that it is less likely to establish an immediate chain of continuous contact with the blood of its victims. Safe medical care and burial practices also help break the chain of transmission, which helps stop the disease from spreading.

On the other hand, malaria, a vector-borne disease, has evolved a stable strategy for infection. The malaria parasite, *Plasmodium*, reproduces in mosquitoes, allowing for the transmission of the parasite to humans through mosquito bites. As long as the mosquitoes exist and humans remain in contact with them, the chain of transmission of malaria remains unbroken.

The figures focus only on the individuals that are infected in a population. In reality, in the case of both malaria and ebola, there are many other individuals who do not get infected.

Emerging Vector-borne Diseases

VARUNI KULASEKERA

INTRODUCTION

Unlike microbes that are transmitted through aerosols, water, food, or body fluids, vector-borne pathogens require transmission between arthropod vectors and vertebrate hosts. Historically, vector-borne diseases such as malaria, yellow fever, and plague were known to have caused major epidemics throughout the world with devastating effects on human populations. During the past two decades, there has been an increase in the incidence of many vector-borne diseases. Emerging vector-borne diseases may be new, resulting from genetic changes in the already existing microbes, or already existing diseases may invade new geographical areas or susceptible host populations, causing a reemergence.

Many factors may be responsible for the emergence and reemergence of vector-borne diseases. Most result from human interventions. Some of the major factors are:

- climatic changes such as increased temperatures and heavy rainfall resulting in the increase of vector populations;
- large-scale irrigation projects, deforestation, and mining which alter the habitats of disease vectors and increase human contact with the vectors;
- overcrowding due to migration, population growth, civil strife, and poverty resulting in poor sanitation and an unsafe water supply, which create ideal habitats for the vectors and increase the probability of human-vector contact;
- trade and travel resulting in exposure to new diseases as well as global spread of disease;
- misuse of insecticide applications creating insecticide resistance among vectors;
- development of drug resistance in pathogens;
- lack of organized and efficient public health programs for disease control and prevention;
- the evolution of new pathogenic microorganisms.

Varuni Kulasekera is a Research Scientist in the Department of Entomology at the American Museum of Natural History and a Content Specialist on infectious disease for the National Center for Science Literacy, Education, and Technology.

SOME EMERGING AND RE-EMERGING VECTOR-BORNE DISEASES

MALARIA The parasitic disease malaria is caused by a protozoan in the genus *Plasmodium*, and is transmitted by *Anopheles* mosquitoes. Malaria has made a dramatic comeback during the last two decades. Every year, 300 to 500 million people are infected with malaria worldwide, with reports of approximately two million deaths. With the advent of DDT, it was believed that malaria would be eradicated completely by eliminating the vector mosquito. This was nearly accomplished in Sri Lanka in the early 1960s with massive campaigns of DDT spraying. The jubilance over the complete control of the vector did not last for too long because the mosquitoes started to develop resistance to DDT. In 1969, there was a major epidemic with over 500,000 people infected with malaria as opposed to less than twenty in 1962. A similar epidemic occurred in 1976 in India, where seven million people were infected. Such trends are now seen in countries such as Korea, Thailand, Pakistan, the tropical Americas and Africa, the latter accounting for over ninety percent of the reported malaria cases in the world.

Some of the important factors that have contributed to the reemergence of malaria are: 1) insecticide resistance among anopheline mosquitoes, 2) drug resistance in the parasites, 3) mass movement of people into malarious areas or spreading of the disease by people from malarious areas to nonmalarious areas because of agriculture practices, wars, and political disruption, 4) creation of new habitats for the vector because of climatic changes, destruction of forests by activities such as mining, road building, and slash-and-burn agriculture, 5) agricultural practices associated with large-scale irrigation projects that bring a significant number of people together, and increase habitats for the vector, 6) lack of concrete public health policy and economic support for disease prevention, and 7) changes in the behavior of the vector.

LYME DISEASE Discovered in the mid-'70s in the United States, Lyme disease is caused by a tick-borne pathogen, *Borrelia burgdorferi*. Although it was initially reported from the East Coast of the United States, the geographical distribution of the disease is on the increase, as well as the disease incidence. In the early '80s, only a few cases (400) were reported in the United States. However, the number of cases is increasing each year, with over 15,000 reported for 1997. Today, Lyme disease is the number one emerging vector-borne disease in the United States. Continuing climatic changes, other environmental changes, and deforestation are the primary causes for the increase in Lyme disease incidence.

PLAGUE Over the course of the past five centuries, plague has caused major pandemics in the world because of the abundance of both its host, rats, and the vector, a flea (*Xenopsylla cheopis*). The incidence of plague was on the decline during the latter part of this century because of effective programs of vector control, rat control, and antibiotic treatment. Only 200 cases were reported in 1981. However, plague has resurfaced in countries such as Africa and India in the recent years because of insufficient public health measures.

YELLOW FEVER The virus that causes yellow fever is transmitted by a species of *Aedes* in Africa, South America, and the Caribbean Islands, and *Heamogogus* in South America. In general, yellow fever transmission occurs in forest habitats where the reservoir host is monkeys. Infected mosquitoes such as *Aedes africanus* and *Heamogogus* maintain jungle yellow fever transmission from monkey to monkey. Other *Aedes* species that are in contact with humans and breed in the periphery of the forest transmit the disease directly to humans causing sporadic outbreaks. Large epidemics of yellow fever occur when the virus is introduced into urban areas where the population is highly susceptible to the disease. In this case, *Aedes aegypti* transmits the virus from person to person.

Although major epidemics of yellow fever have occurred during the last four centuries in Africa and the Americas, the disease was effectively controlled in the Americas during the mid-twentieth century. This was mainly achieved through a combination of successful vaccination programs and eliminating breeding sites of the vector. Today, yellow fever has reemerged in the tropical areas of Africa and the Americas with 200,000 estimated cases and about 30,000 deaths per year. The incidence of yellow fever has increased as a result of breakdown in public health measures (especially the vaccination program), deforestation, urbanization, explosion of the urban vector population and travel.

DENGUE FEVER Dengue fever and dengue hemorrhagic fever occur in the tropical areas of Asia, the Pacific, and the Americas. Evidence suggests that the dengue virus evolved in Asian forest habitats, circulating in a forest cycle involving mosquitoes as their vectors, and monkeys as the reservoirs. The virus is still found in both Asian and West African forest habitats. However, dengue has mainly become an urban disease in the recent years. The virus is transmitted to humans by mosquitoes in the genus *Aedes*, the most common being *Aedes aegypti*.

In the past two decades, dengue fever and dengue hemorrhagic fever have become a major public health concern in many parts of the world, with increasing mortality among humans due to dengue hemorrhagic fever. In the last ten years, the incidence of dengue has dramatically increased with approximately fifty to one hundred million cases of dengue fever and 500,000 cases of dengue hemorrhagic fever reported each year. There are several factors responsible for the increase in disease incidence. An increase in the population growth in developing countries and urbanization associated with it have resulted in poor water supply and sanitation measures. This in turn has increased the population of the vector mosquito which breeds in places associated with urban dwellings, such as water containers, used plastic cans, and used tires. Increased movement by people (mainly air travel) has also resulted in spreading the mosquito and the virus, thus increasing the global spread of dengue fever and dengue hemorrhagic fever.

RIFT VALLEY FEVER For many years, Rift Valley fever was known to affect mainly livestock in Kenya and other sub-Saharan countries. The Rift Valley fever virus is transmitted among livestock by some *Culex* and *Aedes* mosquitoes. The construction of the Aswan Dam in Egypt in 1977 caused an emergence of Rift Valley fever with a dramatic switch from livestock to humans, infecting over 100,000 people and killing thousands. The most recent outbreaks occurred in South Mauritania in 1987, Madagascar in 1990–1991, and Northern Kenya and Somalia in 1997, with similar effects.

This is a classic example where the pathogen has shifted from its original host, livestock, to humans. Heavy rainfall and flooding have created ideal breeding grounds for the mosquitoes near human habitations. The abundant vector population is transmitting Rift Valley fever to humans who are in constant contact with the animals.

FACTORS THAT DRIVE HOST-PATHOGEN EVOLUTION

The concept that vector-borne diseases have been emerging and reemerging is not new. The major factors that cause these disease outbreaks have been documented for centuries. However, a frequently overlooked factor has been the impact of microbial ecology and evolution in the emergence of new disease. The following ecological and evolutionary concepts help explain how and why pathogens have evolved to exploit diverse habitats.

Host and vector are critical for the survival and reproduction of the pathogen. An ideal host and a vector should be abundant, remain infectious for long periods of time, and maintain contact with each other. Pathogens manage to maintain a tight association with specific hosts and vectors as a survival mechanism. Therefore, host and vector specificity is an important factor for their transmission and survival.

During their life cycle, vector-borne pathogens migrate between the vertebrate host and the vector. The pathogen takes great risks in moving from one environment to the other to infect new hosts. They are adapted to reduce these risks by having a high reproductive rate and a short generation time. A single vector can infect several vertebrate hosts, and at the same time pathogens are ingested from a single vertebrate host by many vectors. The pathogen has a greater chance to infect new hosts much more efficiently, thus increasing its overall success during the transmission cycle.

When the vectors are highly diverse the pathogen's opportunity for diversification by occupying different niches will also increase. In North America, eastern equine encephalomyelitis (EEE) virus is transmitted by the mosquito *Culiseta melanura* which is placed in a genus with a few species (thirty-three species). In contrast, Venezuelan equine encephalitis (VEE) is transmitted in Central and South America by mosquitoes in the subgenus *Melanoconion* (genus *Culex*) which is highly diverse (159 species). Like their species-rich vectors, the VEE viruses are also highly diverse, whereas the EEE virus is not.

The population size of the host and their geographical distribution are also important for a pathogen to colonize new hosts. If the host population is large and widely distributed, a pathogen's chances of reaching a new host increases remarkably. For example, the rate of dispersal of EEE virus is very high because it is associated with migratory song birds. In contrast, the rate of dispersal is limited in VEE virus because it is associated with small sedentary forest-dwelling rodents. This increases the likelihood of extinction of VEE viruses.

In vector-borne pathogens, speciation may occur because of vector or host switching. Many pathogens are highly sensitive to even minor changes in their habitats because of their small size and short generation time. When they shift to a different host or a vector, the genetic makeup of the pathogen will change in order to adapt to the new environment. This in turn could create new species of pathogens.

PATHOGENS MUST ADJUST THE DEGREE OF VIRULENCE TO SURVIVE

Pathogens exhibit different levels of virulence towards their hosts, from a mild reaction such as a sore throat associated with the common cold, to severe effects that could ultimately cause death of the host, as in malaria and dengue fever. This contradicts the long-held dogma that host-parasite interactions ultimately evolve into a benign relationship because the survival of the host increases the pathogen's chance of survival. However, if the host dies, the pathogen's survival can reduce immensely, ultimately leading to its own extinction. To overcome this, highly virulent pathogens have increased their chance of survival by increasing their rate of reproduction, maintaining a high level of transmission and using multiple hosts.

This type of increased virulence is associated with vector-borne pathogens. Pathogens that are transmitted by mosquito vectors, such as the malaria parasite and dengue and yellow fever viruses, are highly virulent and often cause death of the host. They do not have to depend on the mobility of the vertebrate host for transmission because the vectors are adapted to disperse the pathogens efficiently. The pathogen has evolved to multiply within the mosquito and the vertebrate host, thus increasing their reproductive efficiency.

THE FUTURE AND VECTOR-BORNE DISEASES

Historically, pathogenic microbes and their vectors were confined to forest habitats. However, with the increase in human population in recent years, activities such as agriculture and irrigation development, and urbanization have also escalated. The combined effects of these activities designed to improve the standards of living for many people around the world have had a major negative impact on the environment, causing the spread of vector-borne disease.

This constant change in the global environment has led us to expect a pattern in the increase of emerging vector-borne infections in the present and future. The number of diseases, such as malaria, Rift Valley fever, and arthropod-borne virus infections, is likely to increase in the future because of large-scale irrigation projects and the mass movement of people. Current development projects in Papua New Guinea that involve forest clearing are likely to bring susceptible human hosts into contact with unidentified pathogens and their vectors. Although yellow fever has never been reported from Asia, there is a likelihood that it might become established in this region because of the availability of the host and the vector. Yellow fever was eradicated from North America after the elimination of the vector breeding sites. However, during the last decade *Aedes aegypti*, the vector of yellow fever and dengue fever, has been invading North America. We have to be alert to the possibility of these diseases emerging in North America.

An understanding of the evolutionary processes of the pathogens and their vectors is imperative if we are to make advances toward conquering these virulent diseases. For example, prediction of the severity of newly emerging pathogens can assist in the selection of different treatment methods for that disease or in planning preventive measures. Also, highly virulent pathogens can be transformed into mild ones. Future decisions in public health policy need to accommodate the findings from evolutionary studies that would help to predict new and emerging disease.

How is a vector-borne disease different from a disease which is transmitted via microbes?

In the treatment of vector-borne diseases such as dengue fever, what treatments might be effective?

Why are treatments for vector-borne diseases different from those used in treating other diseases?

During the Industrial Revolution of the nineteenth century, a mysterious and horrible disease periodically raged through London. People were seized by violent diarrhea, which poured from them in an odorless watery flood containing small white particles, vividly described as "rice-water stools." Vomiting and muscle cramps often followed, as did extreme dehydration and loss of essential salts, which led to severely low blood pressure and fatal collapse of the circulatory system. For reasons no one understood, most of the victims were poor residents of overcrowded neighborhoods, where indoor plumbing was unheard of and sewage flowed through open gutters into the river Thames.

The disease was known as cholera, but no one knew what caused it or how it spread. There was no way to treat it, and death came swiftly and surely to its victims.

Cholera was not a new disease. In India, where it is still endemic, a disease resembling cholera was described in ancient texts dating from as early as 500 B.C. In the nineteenth century alone, six cholera epidemics spread westward from the Indian subcontinent to Europe and most of the rest of the world, beginning in 1817.

It was not until 1883 that the German physician and bacteriologist Robert Koch identified the microbe that causes it as *Vibrio comma*. Now called *V. cholerae*, it is a waterborne bacterium that produces a toxin, which in turn activates an enzyme in intestinal cells to produce the characteristic "rice-water stools," consisting of bits of intestinal lining and gallons of fluid pumped by the small intestine from the blood and tissues. Koch is credited with establishing the "germ theory of disease," but the man who pioneered knowledge about cholera was a London doctor named John Snow.

In 1849, Snow published a pamphlet, *On the Mode of Communication of Cholera*, in which he outlined his theory that it was a contagious disease caused by a poison contained in the vomit and feces of victims of the disease and spread, not by airborne particles, but by way of water contaminated by those excretions. Snow's theory was only one of several floating around at that time. Some people thought cholera was spread by the exhalations, or effluvia, of the victims; this camp was called the "effluvyists." A second camp, the "miasmatists," claimed a deadly vapor, or miasma, rising from the garbage- and sewage-flooded gutters of central London, was inhaled by victims, causing the disease. Other totally unscientific theories held that cholera was a punishment from God for the amorality of the poor, its principal victims.

At the time he published his pamphlet, Snow was a well-known surgeon and Queen Victoria's personal physician, but he first became interested in cholera at the age of eighteen, when he was a surgeon's apprentice, the most common route to a medical career at that time. London was in the grip of the second of the century's six cholera epidemics and the young man observed the devastation first-hand.

In 1849, at the height of the third wave of cholera, one specific London neighborhood was especially hard hit. Within a ten-day period, nearly 500 residents of the area intersected by Broad and Cambridge streets were stricken. Dr. Snow visited the homes of victims, interviewing them about their daily habits and, in particular, the source of the water used by the household. He determined that the overwhelming majority drew the water from a public pump on Broad Street. Certain that the water flowing from that pump was contaminated with cholera, Snow convinced local authorities to take the pump out of operation by removing its handle. New cases of cholera dropped precipitously.

Legend has it that this action alone ended the two-year scourge, but in fact the epidemic was waning by then. Still, it was the first reported instance of direct action to end a public health menace based on scientifically gathered epidemiological data. And it confirmed in Snow's mind the means by which cholera spread. It took him another five years and another epidemic wave to prove his theory, but it is remarkable how close he was to the truth in 1849.

During the fourth epidemic (1853–54), Snow returned to the streets and his painstaking detective work. By then, street corner pumps had been replaced by water delivered directly to houses by one of two private water companies. By interviewing residents and correlating cases of cholera with the source of household water, Snow determined that customers of the Southwark and Vauxhall water company were coming down with cholera at a rate many times greater than those who bought their water from the Lambeth company (1263 versus 98).

Snow also discovered that the Lambeth company water was taken from an area of the Thames that was upstream of where city sewage emptied into the river, whereas the Southwark and Vauxhall water was drawn at a downstream location, where it was regularly contaminated with cholera-bearing waste. Snow had proven his theory.

Today, municipal water supplies are treated by filtering and the addition of chlorine to remove the threat of cholera and other waterborne pathogens, thanks in large part to John Snow's medical geography. But even though we now know how to prevent cholera outbreaks and how to treat the disease when it does occur, it remains a threat to public health in many areas of the world. Clean water and sanitary disposal of human waste are the keys to prevention; replacement of lost fluids and salts, by mouth or in severe cases intravenously, lowers the risk of death to one percent; and antibiotics shorten the course of the disease. But in many developing countries and in times of war and natural disaster, these systems may be lacking and medical facilities and supplies may be hard to come by.

The technology for tracking the incidence of cholera has changed dramatically since John Snow mapped a small London neighborhood using little more than shoe leather and his own analytical skills. Because *V. cholerae* lives symbiotically with certain aquatic plant and animal organisms, satellite imaging equipment is used to track the movement of specific marine colonies associated with cholera outbreaks in coastal areas.

Scientists combine knowledge from such diverse disciplines as ecology, oceanography, marine biology, microbiology, medicine, epidemiology, and space science to predict where cholera will strike next. These predictions may help prevent the disease from reaching epidemic proportions in that area. By giving public health authorities advance warning, preventive measures can be instituted and adequate medical supplies made available. Compared to this, Snow's methods were primitive indeed, but his story remains one of the best examples of how the study of environmental factors enhances the understanding of infectious disease.

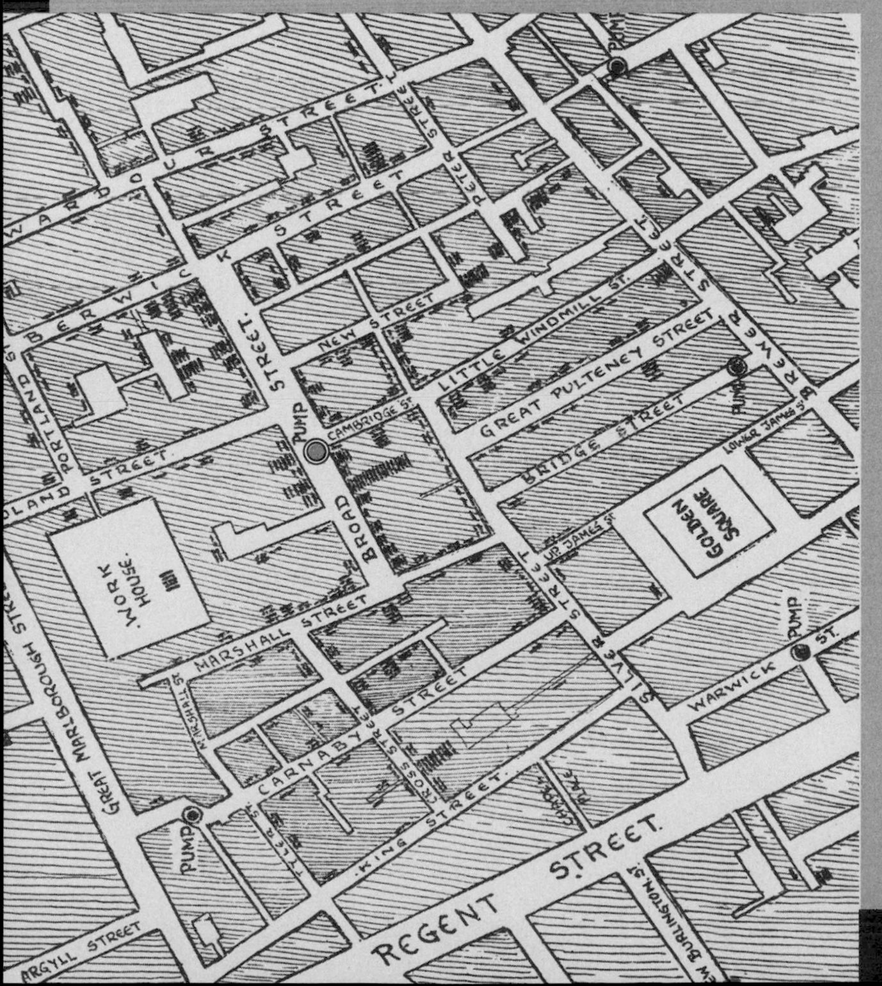

Map of London, circa 1850, showing where John Snow determined the presence of cholera.

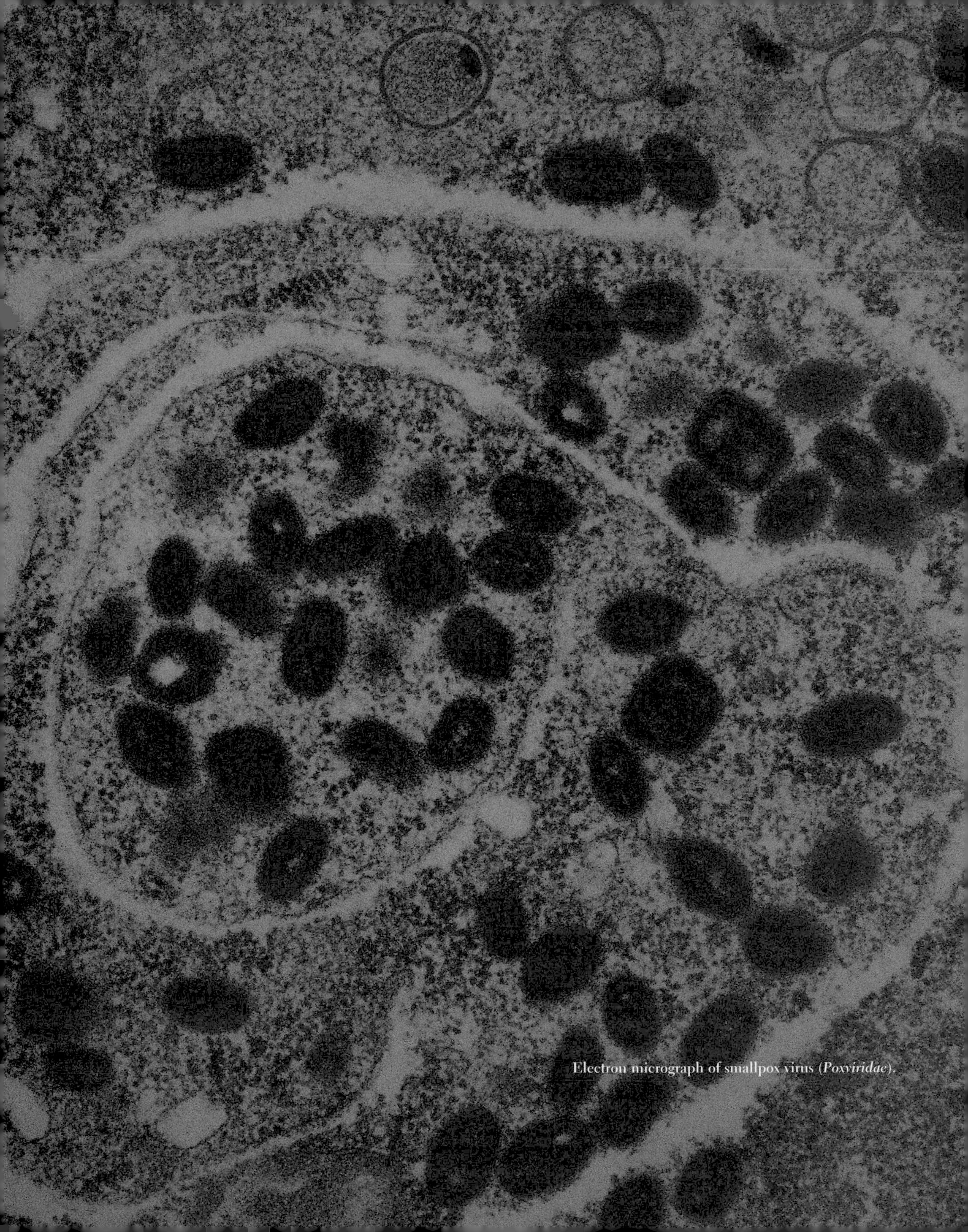

Electron micrograph of smallpox virus (*Poxviridae*).

Diagnosing and Managing Infectious Diseases

JEREMIAH A. BARONDESS

INTRODUCTION

Although infectious diseases are only part of the full spectrum of diseases, they are nevertheless very widespread. In western societies, for example, they are the commonest disorders seen by physicians in non-hospitalized patients, and in developing nations they are the leading causes of disability and death, especially in infants and children. Consequently, the diagnosis, treatment, and prevention of infectious disease are—and will probably always remain—a major component of the effort to preserve human life and increase human happiness.

DIAGNOSING INFECTIOUS DISEASES

For the physician, diagnosing a patient's illness is a kind of detective work that begins with questioning the patient. Typically, about eighty percent of the information needed for a diagnosis—the determination of the specific nature and circumstance of the disease—comes from the patient's history.

The physician faces maximum uncertainty upon beginning this initial questioning: the patient could have any one of a vast number of diseases. Consider, for example, the case of a 17-year-old girl who shows up in her physician's office with a fever, and reports that she has had it for three weeks—an alarming general symptom. The first questions the physician asks her are guided by the four general causes of fevers: infections, tumors, exposure to harmful substances, and "inflammatory diseases" such as lupus or rheumatoid arthritis. The physician then makes a review of her body systems by asking a series of questions to learn whether she has had any focal symptoms—symptoms that point to the organ or organ system where the infection or other problem is. For example, has she had headache, cough, or sputum, symptoms suggesting a respiratory system infection? Has she had symptoms suggesting a urinary tract infection? If her answers do not point to a focus of the illness, this

Jeremiah A. Barondess is President of the New York Academy of Medicine and Professor Emeritus of Clinical Medicine at the Cornell University Medical College.

suggests something might be seriously wrong in an internal organ, such as a pelvic abscess, bacterial endocarditis, cancer, or leukemia, among other serious possibilities.

After reflecting on the answers to the initial questioning, the physician must use further narrowing-down techniques to minimize the level of uncertainty in the diagnosis. This uncertainty is never zero; at best, a diagnosis can reach a high level of probability. Consider the case of the 17-year-old girl whose initial history was just described. Her list of possible diagnoses would be narrowed down by asking her questions about her health background: What health problems, if any, have her immediate family members had? Does she have any long-term health complications? What diseases might she have been exposed to? Has she been exposed to sick people? Has she recently traveled in a part of the world where infectious diseases are prevalent? Has the disease taken a toll on her overall health? Has she lost a significant amount of weight? While carrying out this line of questioning, the physician formulates diagnostic hypotheses and tests them against the information provided by the patient.

After obtaining the patient's history, the physician further narrows the list of diagnostic possibilities by performing a physical examination—which typically provides another 10 percent or so of the information needed for a diagnosis. For example, a tender mass in a patient's abdomen might indicate an abscess; an enlarged liver might indicate hepatitis or a tumor.

The final ten percent or so of the information required for a final diagnosis involves specific tests which are chosen to decide among the remaining diagnostic possibilities. Such tests commonly include the blood count and the urinalysis. Increase in the number of white blood cells is a common feature of infections: in bacterial disease, it is the polymorphonuclear leukocytes that often become more numerous; in viral infections, an increase in the proportion of lymphocytes (which make antibodies) may occur. The presence of white blood cells and bacteria in the urine may indicate infection of the genitourinary tract. Additional common tests include cultures of blood, urine, sputum, spinal fluid, pus, or skin lesions. These culture techniques can result in isolation and identification of the offending organism if it is a bacterium and can also provide an opportunity to test its sensitivity to various antimicrobial drugs. Identifying viruses by culture or other techniques is more difficult, takes longer, and is often less useful than for identifying bacteria—because common viral diseases can often be diagnosed with reasonable confidence without isolating the virus. Specialized techniques sometimes used in the diagnosis of viral infections include the polymerase chain reaction (PCR) to multiply viral DNA in the lab so the virus can be identified, electron microscopy to identify the virus from its appearance, and, most commonly, tests for antibodies in the blood to specific viruses. The identification of antibodies may also be useful for infections caused by bacteria and protozoans such as amoebas.

Although expensive, body-imaging techniques such as X-rays, computerized axial tomography (CAT scan), and magnetic resonance imaging (MRI) are sometimes necessary for the diagnosis of infections. An example of this was a patient who had been treated for two weeks with antibiotics for what was initially diagnosed as pneumonia. His symptom was painful breathing (but he lacked the cough characteristic of pneumonia). Despite the antibiotics, he was getting steadily sicker. Physical examination and chest X-ray revealed signs of fluid accumulation in his chest; he had an enlarged liver,

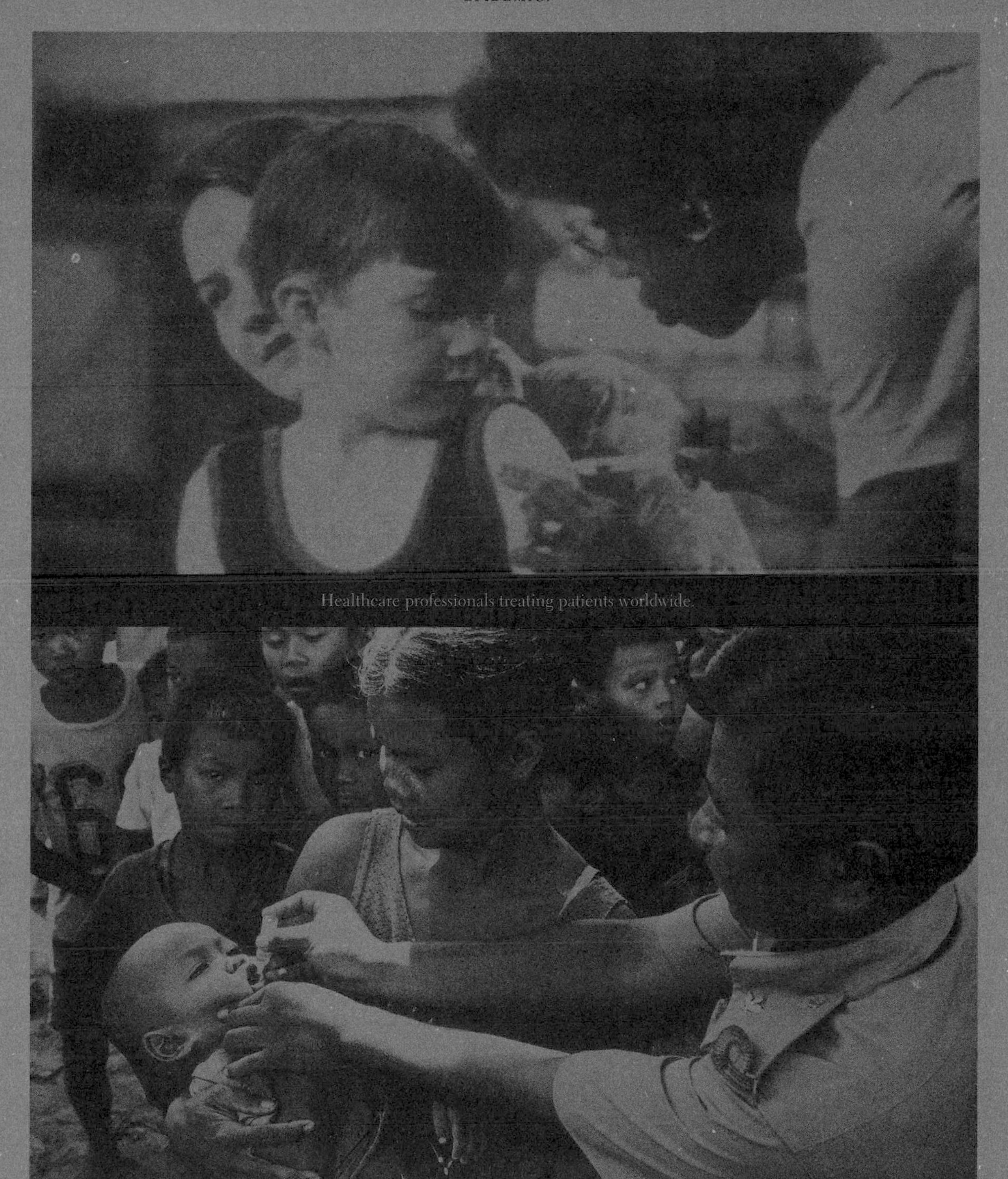

Healthcare professionals treating patients worldwide.

which, together with the information that he had had diarrhea during a recent visit to Puerto Rico, led to the tentative diagnosis of amebiasis of the liver with perforation of the diaphragm so that fluid accumulated in the chest cavity. Fluid removed from his chest and examined under a microscope revealed the presence of this pathogen. This last test was not only necessary, it was specifically diagnostic—it completely specified the patient's disease.

Sometimes the ultimate diagnostic test is to wait and watch—to observe the course followed by the unfolding disease. Most people do not need treatment so urgently that this cannot be done, when necessary. Moreover, if infected people receive antibiotics before the cause of their infection has been determined, their diagnosis could easily be confused.

TREATING INFECTIOUS DISEASES

Whenever possible, infections are treated with antimicrobial drugs known to destroy the identified infecting agent completely. Frequently, particularly when infecting bacteria originate in the gastrointestinal or genitourinary tracts, a mixture of antimicrobial drugs is more effective than a single drug. Although viruses are usually more difficult to treat with drugs than bacteria are, drugs are sometimes effective; for example, in the treatment of a range of herpes infections. Drug combinations are often effective in reducing the number of virus particles in HIV infections, sometimes reducing levels of the virus in the blood to the point of undetectability.

Sometimes patients arrive at an emergency room when they are already severely, even overwhelmingly, infected. In these cases, after appropriate samples have been collected from them for culture, antimicrobial therapy must be started immediately, even though the specific organism causing the illness has not yet been identified. Under these circumstances, multiple drugs are commonly used. The drugs chosen are those with proven effectiveness against the likeliest infecting agent(s)—usually bacteria. Typically, the treatment of infections also involves general physiological support, including bed rest, fluids (administered intravenously, if necessary), careful attention to diet, and frequent reexamination by the physician to learn whether the disorder is responding to treatment and whether any complications have arisen.

THE CHANGING ECOLOGY OF INFECTIOUS DISEASE: A DIAGNOSTIC CHALLENGE FOR PHYSICIANS

Around the turn of the century, a great doctor named William Osler observed that it is not only important which disease the patient has, it is also important which patient the disease has. Osler had in mind that disease represents a balance of factors such as host susceptibility or resistance and the virulence or dose of an infecting organism. To be effective in the fight against infectious diseases, physicians must also be aware of the ecology of infection—an ecology that is constantly changing. As humans increasingly encroach on the habitats of wild animals, "new" infectious diseases, such as the Ebola virus, emerge from populations of non-human vertebrates and infect humans. Modern methods of transportation such as jet aircraft allow new infections to spread throughout the world's population very quickly, so it is crucial for health organizations to monitor infections throughout the world regularly, and for physicians to keep up with the results of this monitoring. On the other hand, when diagnosing, the physician should consider diseases common in the patient's locality and subpopulation before considering exotic diseases from distant locations.

Ecological pressures that have changed the balance between man and bacteria have been caused by the indiscriminate use of antibiotics and have resulted in the emergence of bacteria resistant to formerly effective drugs. This is an important and increasing problem—infections that could formerly be effectively treated are now, in many cases, difficult to manage. In this instance, inappropriate use of antibacterial drugs has had the paradoxical effect of causing a serious problem that can be met by using these medications only in instances in which they are clearly needed and are appropriate to the patient's infection.

HELPING TO PREVENT INFECTIOUS DISEASES

The ideal management of infectious diseases is prevention. On the community level, improving the quality of the water supply has been a major factor in reducing the incidence of cholera, typhoid fever, and other water-borne infections. Effective sewage systems have reduced the transmission of diseases spread by the feces-to-mouth route and careful attention to the quality of the food supply has reduced bacterial infection. We must not become complacent, however, because occasionally major water-borne or food-borne outbreaks of infection still occur. For example, there was an epidemic of cryptosporidiosis that infected 400,000 people, and killed more than 50 of them, in Milwaukee in 1993—and the same year four children died in the Pacific Northwest after eating hamburgers contaminated with *E. Coli* 0157:H7. Moreover, we as a nation need to establish health care policies such that every individual in our society can affordably and conveniently receive the type of care that can reduce the risk of infectious disease. This includes making it convenient and affordable for parents to have their children immunized against the common childhood pathogens by the age of two. Further, despite their political unpopularity, needle-exchange programs could help reduce the spread of AIDS among the population that uses intravenous drugs, such as heroin.

Immunization procedures have eradicated smallpox completely, and have reduced the incidence of poliomyelitis (polio) to the point at which its complete eradication is a realistic prospect. Today, state laws require children to be immunized against measles, mumps, and pertussis before they can start school, but too many have been needlessly infected before then—infections preventable through immunization.

People can greatly reduce the risk of AIDS and other sexually transmitted diseases by practicing "safe sex." On the individual level, even something as simple as paying proper attention to cuts and bruises—soap and water is an extremely effective antibacterial combination—can help protect people from a variety of infections. Nevertheless, unless people spend their lives in hermetically sealed environments, they will always be at some risk of infection. Infectious disease is, and will continue to be, part of the human condition.

Why is a patient's medical history so important in the diagnosis and treatment of various diseases?

What options does a physician have when prescribing treatment for infectious disease?

How can physicians have an impact on disease prevention?

Profile

Mary Wilson: Taking a Global View of Infectious Disease

Mary Wilson, M.D., is an expert on the geographic distribution of infectious diseases. She teaches in the department of Population and International Health and in the department of Epidemiology in the Harvard School of Public Health, as well as in the Center for Health and the Global Environment at Harvard Medical School. Mary Wilson is the author of a 769-page textbook called *A World Guide to Infections,* in addition to sections in several other books and dozens of articles on emerging infections and the role of travel in their spread. But she almost spent her life as an English teacher.

As she tells it, she had been interested in science since childhood and wanted to become a pathologist, but in the 1950s and early 1960s, women became teachers or nurses, and getting an M.D. seemed out of reach.

She was on her way to earning a master's degree in English literature when she decided to reach for a career in medicine anyway. She started taking the science courses required of medical school applicants and spoke with the director of admissions at the University of Wisconsin Medical School, who was extremely discouraging about her prospects. "He told me I was a terrible candidate. Wasn't it obvious that I had done something else before medical school and hence could not be serious? I was a woman and would occupy a valuable space that a man deserved to take."

It is fortunate for medicine that Mary Wilson persisted and gained admission, earning her M.D. in 1971. And it is fortunate for others—male and female alike—that such attitudes have changed enormously in the intervening decades. Not only do women now represent more than fifty percent of those entering medical schools, but admissions officers consider applicants of either sex with academic and life experiences in other fields to be better bets than those with a narrower focus.

Dr. Wilson became interested in infectious diseases while still a medical student. "It had everything needed to provide intellectual excitement and unending challenges. Infections are everywhere; often preventable and curable, involving every part of the body and every age, appearing sporadically and in outbreaks, and often cloaked in mystery that requires careful sleuthing to diagnose or to discover a source. They range from the mundane to the exotic and have shaped and will continue to shape cultures and societies."

It was during a fellowship in Haiti, where she treated patients with such infectious diseases as malaria, typhoid fever, and leptospirosis, that she began asking the questions that have shaped the rest of her professional life. Why, she wondered, were the diseases in Haiti, a country geographically close to the United States, so very different from the diseases seen at home? Why are specific diseases where they are? And how do they spread?

Answering these questions requires an understanding not only of the basic biology of pathogens and their means of transmission, but also their ecology. "For diseases that are spread by vectors or that have an animal as an intermediate or reservoir host," she explains, "environmental conditions may determine whether or not the disease can be maintained and spread in a particular region." Some infections—many tropical diseases carried by insects, for example—are relatively fixed in distribution, whereas others—such as HIV, which is carried by humans—can be introduced into populations throughout the world.

Dr. Wilson sees travel as one of the most significant factors in the spread of disease, both in the past and the present. "Today the volume and speed of travel are unprecedented, and that is likely to increase in the decades to come. Further, worldwide commerce involving the transport of plants, animals, food, insects, and seeds around the globe has given wings to species that would otherwise be confined to one geographic area," she says.

But human travel is only part of it. "Massive population shifts are occurring because of social and political unrest, economic and environmental pressures, and war. Extreme weather events displace people from their homes; refugee camps and temporary settlements often provide an environment where

Mary Wilson

risk of infection is high because of crowding and poor sanitation."

Humans help spread not only disease organisms, but also the means by which increased virulence and microbial resistance develop. "We carry microbial genetic material in and on our bodies that through transfer, recombination, conjugation, reassortment, and a variety of other molecular maneuvers can confer virulence and resistance to other related or unrelated microbes," she explains.

This perspective is one she developed as a member of a multidisciplinary group at the Harvard School of Public Health that was formed with the goal of exploring the factors behind accelerating changes in infectious disease. People in fields such as epidemiology, ecology, entomology, infectious diseases, population biology, evolutionary biology, marine ecology, and mathematical modeling shared their knowledge and experience. "We tried to go beyond our traditional disciplines to think broadly about why we are seeing global changes." This sort of barrier-breaking inquiry is what happens when a young woman who studied French and English literature and philosophy in college is given a chance to pursue her passion for a career in medicine.

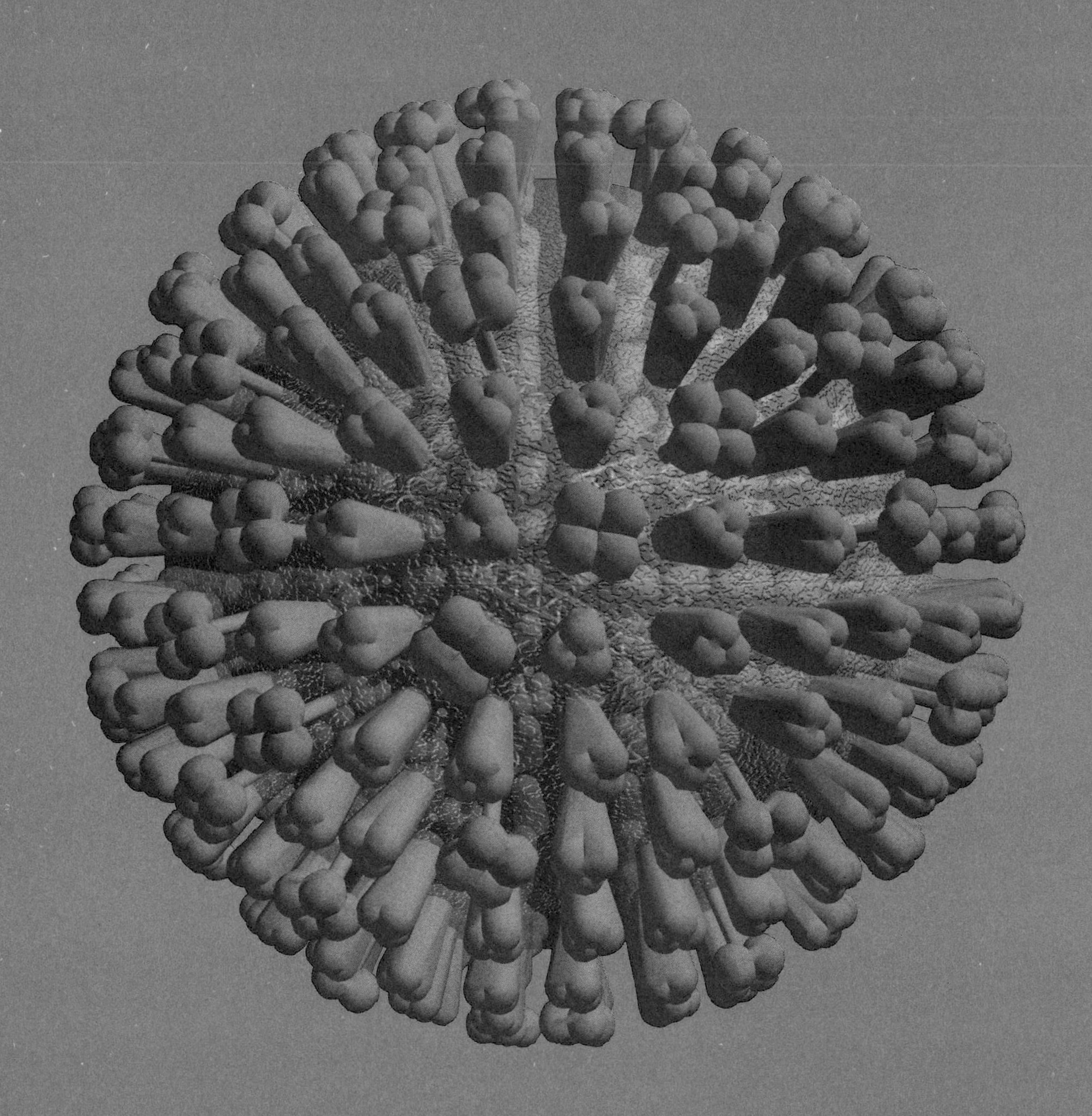

Computer-generated image of a flu virus (Influenza A)

Section Three: Infection

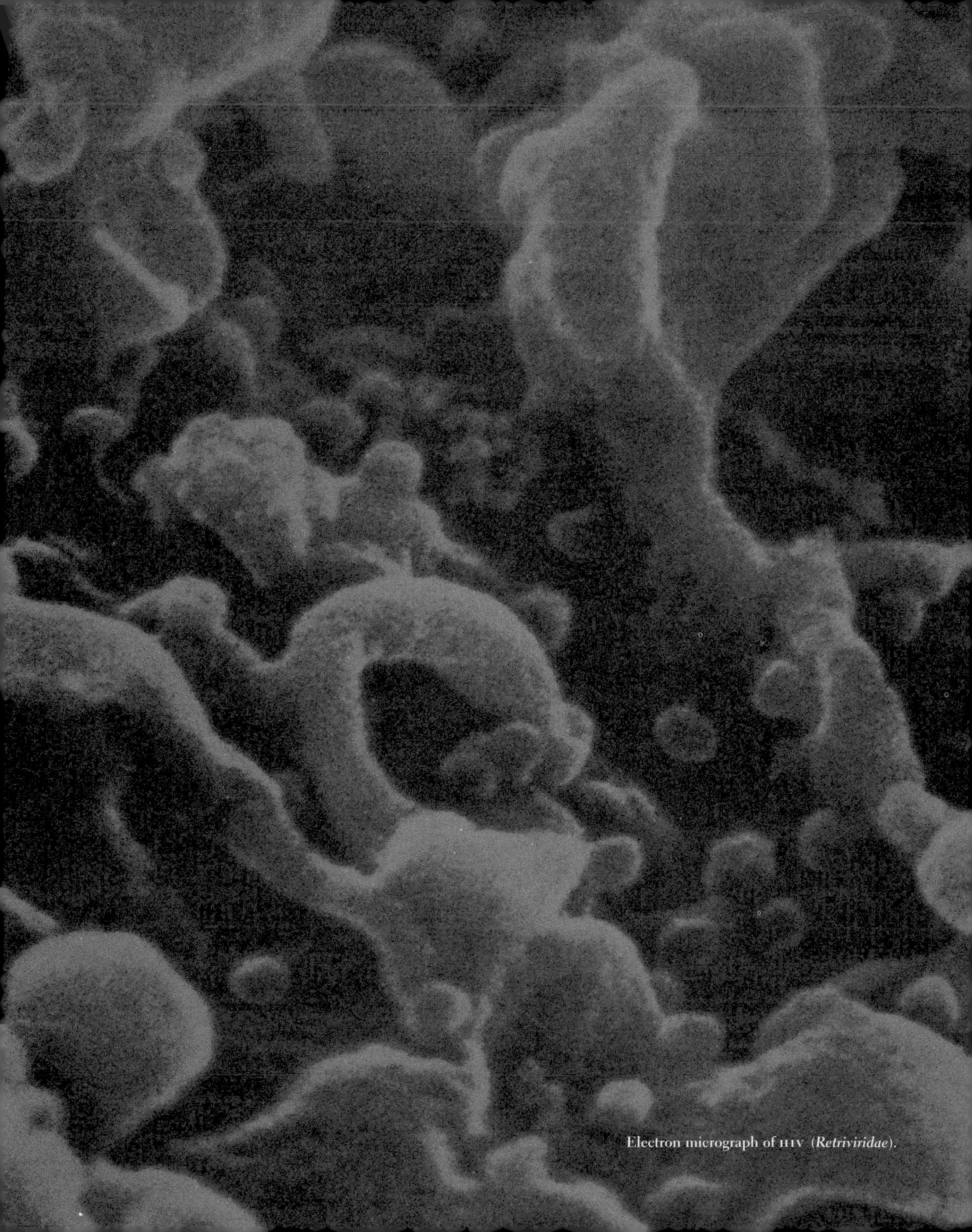

Electron micrograph of HIV (*Retriviridae*).

Section Three: Infection

Introduction

ROB DESALLE AND MARLA JO BRICKMAN

Infection is the state or condition in which a pathogenic agent invades the body or part of it and then, if not checked by the body's defenses, multiplies and produces harmful effects. Most textbooks suggest that infection is "When microbes overcome humans." To do so, they must first enter the human body. This section describes how microbes get into our bodies and how our bodies respond—either to fend off the microbes or to succumb to them. The concentric circle therefore widens in this chapter to include the human body and its defense system.

This section addresses several major aspects of infection: (1) how microbes enter the human body (portals of entry); (2) how the human body responds to infection; (3) how humans have supplemented their bodies' natural defenses against infectious disease through the development and application of antimicrobial drugs and vaccines.

The entry of microbes into the human body can occur only in specific areas of the body termed portals of entry. The most common "natural" portals of entry are the mouth, the nasal passage, the genitals, the skin, and mucosal tissues such as the area around the eyes. Microbes also exploit such "artificial" portals as cuts, skin scrapes, scratches, and sores. The sequence of events that occurs after microbes enter the body is the essence of the disease, producing the "history" of symptoms used to identify the disease. Understanding this sequence requires understanding how microbes attach to human tissue; how they multiply; how the human immune system responds to them; how they evade, or attempt to evade, the response of the human immune system; and what they do to human cells and to the human body overall to cause such symptoms as fever, rash, sores, and inflammation.

As a microbe attempts to enter a human body, it faces an imposing array of defenses. The body's first line of defense to microbes comprises such "nonspecific defenses" (defenses not targeted at a specific microbe but acting against all microbes) as our skin, tears, and mucus. Our skin keeps most microbes out of our bodies, and its defensive effectiveness is maintained by the secretion of oils that protect it from microbial assault. In addition, our bodies have structures that mechanically hamper microbes from exploiting some portals of entry; for example, our noses have small hairs that impede the entry of microbes through the nasal passage. Our normal body chemistry is also part of the defensive line. For instance, chemicals in our tears destroy bacteria, which prevents them from entering the body through the mucosal tissue around the eyes.

Our body's second line of defense against invading microbes is its incredibly sophisticated immune system. We call this system a "specific" defense system because it acts against specific microbes. The immune system "knows" what cells belong to us and can recognize millions of kinds of foreign cells, ranging from disease-causing bacteria to the cells of life-saving transplanted organs. Once the immune system detects a foreign cell, it acts like a massive communication network, directing its various cells where to go and what to do.

When the foreign cell is a microbe, the immune system's "directive" varies depending on the type of microbe and where it has infected the body. Targeting its response to the specific type of microbe, the immune system activates only those immune system cells that can rid the body of that microbe. The immune system then "remembers" the surface properties of that microbe, so that the next time it enters our bodies, it will encounter a faster and stronger immune response. Many types of cells that participate in the body's immune response derive from bone marrow (the source of the B in B cells), although some mature in other parts of the body, especially the thymus (the source of the T in T cells). These cells may act in "cellular" (or "cell-mediated") immune defenses, in "humoral" immune defenses, or in both. Cellular immunity involves the direct interaction of the invading microbes with T cells. Humoral immunity involves the direct interaction of the invaders with antibodies that specifically target them.

When a microbe enters the body, macrophages (large white blood cells) are one of the first types of immune system cells on the scene. They devour invading microbes and other "unwanted" particles in a process called phagocytosis. By placing pieces of the devoured microbes on their surface, the macrophages also activate other types of immune system cells.

There are millions of different variants of a type of white blood cell called T-helper cells. Each type of T-helper cell has a region on its surface that can bind only to one specific type of foreign protein called an antigen. Consequently, when diverse T-helper cells encounter macrophages with a specific antigen on their surfaces, only those T-helper cells specific for that particular antigen will physically attach to it, and thereby to the macrophages. Once attached to a macrophage, the T-helper cell sends out chemical signals (specific chemicals that travel through the body) that call cells of either the cellular immune system or the humoral immune system into action.

The chemical signals that the activated T-helper cells send out stimulate the production of more of the same specific type of T-helper cells. These same "chemical messengers" also activate T-cytotoxic cells (also called killer T cells or repressor T cells); these are white blood cells belonging to the cellular immune system. Unlike T-helper cells, which can only "tag" foreign cells that need to be killed and signal for help, T-cytotoxic cells can destroy invading microbes directly by killing cells they have entered. The signals that T-helper cells send out also activate humoral immune system cells called B cells. As with the T-helper cells, we have millions of different variants of B cells, each with a region on its surface that can "recognize" only one type of foreign antigen. Once the B cells specific for that particular antigen, and therefore for that particular microbe, are activated, they transform into specific types of "plasma cells" that produce and secrete antibodies specific to the antigen. Antibodies are soluble proteins that can neutralize a specific foreign microbe by binding to it. This greatly hampers the microbe often by preventing it from entering its target cells or by causing it to precipitate from the blood to be devoured by a macrophage and/or targets it for destruction by T-cytotoxic cells.

Perhaps unique among living things, humans have developed medicines and cultural practices that supplement their bodies' evolved defenses against infectious disease. These developments include antibiotics and vaccines, and such "low tech" interventions as hand washing using soap, mosquito netting, building houses above heights frequented by insect vectors, and using appropriate contraceptives.

In 1901, Paul Ehrlich coined the term magic bullet to refer to agents that could control and kill disease-causing microbes. He used the arsenic-containing compound arsphenamine (sold commercially as Salvarsan) to treat the previously incurable disease syphilis and a related spectrum of tropical diseases including yaws. Salvarsan was the first in a series of successful artificial "antibacterials"; subsequent successful antibacterials included the sulfa drugs. (Historically, medicine has exploited a handful of traditional natural antimicrobials, including the antimalarial drug quinine.) Soon a much more versatile class of antibacterial drugs, with fewer side effects, appeared: the antibiotics. The term antibiotic originally referred to any chemical agent produced by one organism that inhibits the growth of or destroys microorganisms; now the term also embraces synthetic modifications of naturally occurring antibiotics. In 1928, Alexander Fleming found that *Penicillium notatum*, a green mold, produced a substance that was lethal to many kinds of pathogenic bacteria. A decade later, Ernst Chain and Howard Florey isolated the antibiotic substance—penicillin—and proved its potency and lack of toxicity. Pharmaceutical companies (including Squibb, Merck, and Pfizer) invested time, energy, and capital to develop techniques for producing penicillin in large quantities, with striking success. By 1944, in the latter phase of World War II, adequate amounts of penicillin were available to save many lives of thousands of military personnel who otherwise would have died of massive infection resulting from battle wounds.

At first, penicillin and subsequently isolated antibiotics were hailed as the "miracle drugs" that humanity had long prayed for. However, it soon became apparent that bacteria can evolve resistance to antibiotics. Often, for example, bacteria will continue to grow in a culture medium containing an antibiotic. How have bacteria that once succumbed to a particular antibiotic become resistant to it? Bacteria can evolve such that they lose their vulnerability to the antibiotic, say by modifying their cell walls so that the antibiotic can no longer enter, by changing a metabolic pathway that the antibiotic blocked, or by modifying the specific molecular "target" of the antibiotic. This type of resistance usually results from mutations on the bacteria's chromosomes. Alternatively, bacteria can evolve biochemical pathways that produce substances that either destroy the antibiotic or render it inactive and useless. The genes that "guide" the production of these substances typically occur on a small circular piece of DNA called a plasmid, which the bacteria often received from other bacteria that were resistant to this antibiotic. Some bacteria, such as those belonging to the genus

Staphylococcus (and that cause "staph infections") can "pick up" many plasmids that confer multiple drug resistance (MDR)—resistance to several different antibiotics—on the bacteria. To maintain the potency of today's antibiotics, they must be used intelligently. Unrestrained use will probably result in the rapid evolution of multiple-drug-resistant bacteria. Physicians should prescribe antibiotics only when they are necessary and should ideally choose antibiotics that act against the target bacteria but not against other bacteria present.

For hundreds of years, different cultures in China, India, and Africa have attempted to protect people from smallpox by smearing their skin with, or having them ingest, smallpox lesions from infected individuals. In the early eighteenth century, Europeans adopted the skin-smearing procedure from the Turkish. When the inoculations were successful, the inoculated individuals caught a mild form of smallpox, and then, having recovered, were immune to smallpox. Unfortunately, inoculation sometimes caused the recipients to develop full-fledged smallpox. In 1796, Edward Jenner showed that matter from lesions in cattle or people suffering from cowpox could make people immune to smallpox without causing that disease through the inoculation itself; the names vaccine and vaccination coined for this procedure derived from the Latin name physicians then used for "cowpox," *vaccinia*.

It was only in this century that scientists discovered those workings of the immune system, briefly discussed above, that lay behind the development of immunity through vaccination. Simultaneously, medical scientists developed successful vaccines against many bacteria and viruses. Some of these vaccines consist of pathogenic microbes that have been killed or otherwise deactivated—usually through chemical treatment or heating—to the point that they do not cause serious illness while still causing the body to develop antibodies protective against the pathogen. Some vaccines consist of microbial strains of pathogens that have mutated such that they no longer cause the disease while still causing the body to produce antibodies against the pathogenic strain. A third type of vaccine is directed not against a specific pathogenic microbe, but against the toxin it produces which kills or injures its host. The best known example of this type of vaccine is tetanus toxoid, which provides immunity against the potent neurotoxin, produced by bacillus *Clostridium tetani* that causes the symptoms of tetanus or lockjaw. Tetanus toxoid used as a vaccine is a form of the neurotoxin modified such that it is no longer toxic, instead it is a "toxoid" but still stimulates the production of antibodies against the toxic form. Some vaccines are isolated antigens of microbes; the antibodies the immune system develops against the particular antigens in the vaccine suffice to ensure the destruction of the entire organism should it enter the body.

Vaccination programs are most effective when they reach the largest number of people lacking immunity to a specific pathogen. When this does not happen, the consequences can be severe. Between 1989 and 1991, many children lacking immunity to measles failed to receive vaccination against this disease; as a result, 55,000 of them caught the measles. Many states have enacted public health regulation requiring school-age children to have the prescribed series of vaccinations before they can enter school; programs such as Vaccines for Children provide these vaccines for children unable to obtain them otherwise.

To explore infection, we pose the following questions:

How are infectious diseases spread?

ROBERT SHOPE, Professor, Center for Tropical Diseases, The University of Texas Medical Branch, describes the roles of the major factors in the spread of infectious disease: air, animal bites, vectors, blood and other bodily fluids, food, and water.

How does our immune system fight infectious disease?

BARRY BLOOM, Investigator, Howard Hughes Medical Institute, Albert Einstein Medical Center, discusses the importance of immune system research in order to develop practical treatments of infectious disease.

How are vaccines developed and tested?

ANNE GERSHON, Professor of Pediatrics/Infectious Diseases, Columbia University College of Physicians and Surgeons, focuses on the clinical trials of a chicken-pox vaccine, to show the crucial role that health clinics play in the development of effective vaccines.

How are antibiotics and antiviral drugs developed?

RICHARD COLONNO, Vice President, Infectious Diseases Drug Discovery, Bristol-Myers Squibb Company, documents the process of drug discovery from basic research through development and delivery.

As these essays demonstrate, the study of infection—from understanding how foreign invaders enter the human body to charting the immune system's ability to recognize trespassers—establishes the foundation for the further development of vaccines, antibiotics, and antiviral drugs.

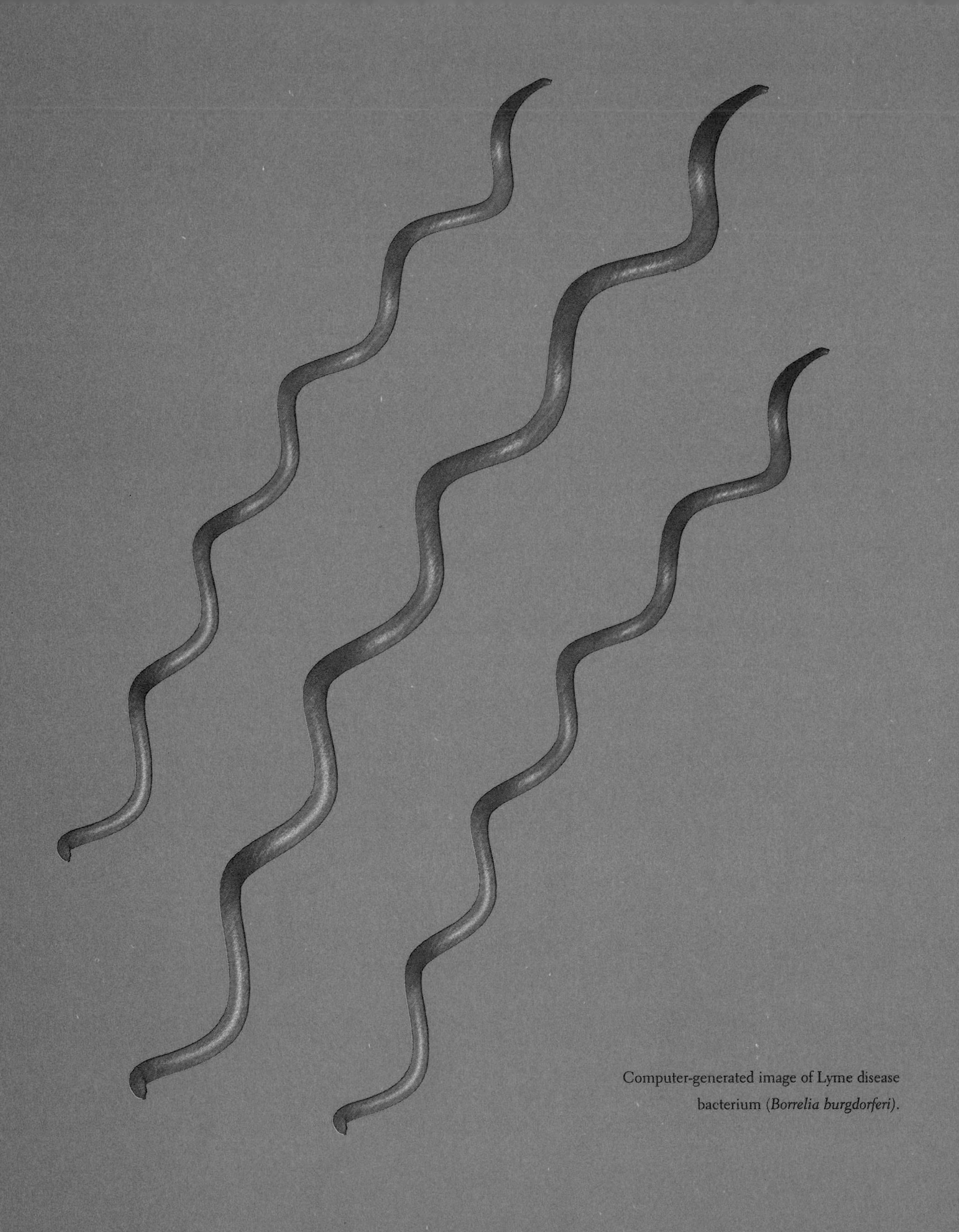

Computer-generated image of Lyme disease bacterium (*Borrelia burgdorferi*).

How Infection Spreads

ROBERT E. SHOPE

TRANSMISSION BY AIR

Kerchoo! That sneeze comes over you suddenly, and all your nearby friends look up. "Bless you," says one of them. You have a cold. What you do not know and what no one can see is that the cold was caused by a virus, a tiny microbe that was multiplying in the lining of your nose and throat. This virus was clever. It seemed to know that it had to get out of your nose and throat and into the air, so that it could spread to your friends. As it multiplied in the cells of your nose, it irritated them and sent your brain a message that you would sneeze. Your sneeze propelled millions of cold virus particles into the air, where other people could breathe them. Micro-droplets of just the right size stayed suspended in the air long enough for some of them to find the noses of your friends. There is truth in the saying "catch a cold." Just as you catch a baseball thrown to you, you catch (in the linings of your throat and nose) the cold virus—which is even shaped like a tiny baseball.

Viruses and other microbes have special coatings with little docking sites. These sites enable the microbe to "dock" with a specific type of cell—and only with that type of cell—in a potential host. Each "target cell" in the host has a receptor with precisely the right shape to fit the docking site on the microbe. When the microbe comes upon a target cell, it docks with it. In this way, the cold virus finds the right cell in your friend's throat or nose, attaches to it, and is invited in. Once inside, it makes more virus, which escapes and starts the disease anew. Other microbes such as influenza viruses behave the same. This means of transmission is called "aerosol transmission."

TRANSMISSION BY ANIMAL BITES

Not all microbes travel from person to person through the air. Over the ages, each species of microbe has evolved adaptations that help it survive; these adaptations include its own ideal way to travel between hosts. Consider the rabies virus. Rabies is a disease that infects domestic dogs and cats, and such wild animals as coyotes, raccoons, and bats. The rabies virus infects the brain and nerves of its victims. Because the virus does not have access to

Robert E. Shope is Professor of Pathology at the Center for Tropical Diseases, University of Texas Medical Branch, Galveston, Texas.

the nose or the lungs of its victims, it cannot spew itself into the air like the virus causing the common cold. It must find a different method of transmission, and it is very clever. It spreads through the bite of an infected animal to another animal.

In many parts of the world, rabid dogs commonly attack and bite people. What triggers this unusual aggression by a rabid dog? The rabies virus travels within the dog's body and establishes an infection, first in the dog's brain, then in its salivary gland. The dog is normally friendly and loves to be with people. However, when the virus has infected its limbic system (the center of the brain that controls its emotions) the dog becomes "mad," losing control and becoming angry, biting anything that comes in its way. The virus in its saliva infects the person's bite wound.

Fortunately, although the rabies virus was very clever to devise this method of moving from victim to victim, scientists were even cleverer. More than 100 years ago, the French physician and scientist Louis Pasteur devised a rabies vaccine for dogs. If you have your pets vaccinated, you no longer need fear that they will be infected with rabies.

TRANSMISSION BY VECTORS

Vector transmission is similar to bite transmission except usually the virus carrier does not get sick from the microbe. Some microbes go to great lengths to survive and maintain themselves in nature. One example of these causes fever, a red rash, and sometimes joint pain and swelling, and even nerve damage. The illness caused by this microbe is called Lyme disease because it was first noticed in Old Lyme, Connecticut, in 1975. The mothers of two young boys with arthritis mentioned to Dr. Allen Steere, a young doctor at Yale University, how strange it was that their children and other neighborhood children had had the same disease simultaneously. Dr. Steere soon found that many other arthritis cases had occurred on the same rural road where the two boys lived. He did not know at the time what caused Lyme disease, but he thought it must be a microbe. Because ticks were plentiful in the neighborhood, they were suspected to be the transmitters of the unknown microbe. Later investigators identified the infectious agent as a species of spiral-shaped bacteria that received the name *Borrelia burgdorferi*. To understand what caused this outbreak of Lyme disease, it is helpful to consider both the natural history of *Borrelia burgdorferi* and what was happening in Connecticut in 1975.

The borrelia infect and live in small birds and mice but do not make them sick. These hosts provide meals of blood to the immature stages of the deer tick, a barely visible tick about the size of the head of a pin. The ticks become infected with the borrelia that arrived in the blood from the infected birds or mice. The borrelia multiply in the mid-gut of the tick and are transmitted when the tick feeds again. When the tick grows into an adult, it feeds on the blood of deer; the deer do not become infected with the borrelia but are very important to the tick as a supply of blood to nourish its eggs. The birds, mice, deer, ticks, and borrelia lived happily together in the Connecticut forests, apparently without hurting anybody. What happened in 1975 to cause the outbreak of Lyme disease?

Consider two events. First, deer became much more common because the forests, which had been cleared during the preceding decades, grew again, providing shelter and food for the deer. The expanded deer population in turn provided food for the ticks, which became much more numerous. Second, people started to build their homes in country areas. The combination of these two factors exposed many more people to the borrelia, especially children who wanted to explore and play in the trees in their backyards. The ticks, which had previously been less common and did not bother people, now fed on the children. In the process, they injected borrelia in their saliva under the children's skin, infecting them. Many other infectious microbes use ticks, mosquitoes, fleas, and biting sand flies as their vectors. However, these vectors do not contract the disease.

TRANSMISSION THROUGH FLUIDS

HIV, the virus that causes AIDS, is not a particularly clever virus like rabies. For its transmission, HIV relies on direct introduction of the virus which is in blood or semen into the body of another person. In particular, its transmission requires close sexual contact, blood transfusion, intravenous injection by drug users using HIV-contaminated needles, or infection of a baby by its mother's blood at or near the time of birth. Intravenous blood injection also spreads the hepatitis B and C viruses. Once HIV or hepatitis virus enters the body, it finds a specific type of target cell. For HIV, the target is a type of immune cell, and for hepatitis B and C, liver cells. After the virus docks with its target, it enters and then produces new viral particles, which can be transmitted in turn. Sexual contact is also responsible for spreading other diseases, including syphilis, gonorrhea, and herpes-2. In these infections, the microbe infects cells of the mucous lining of a person's sex organs, forming sores from which it can spread to another sexual partner. Syphilis can also spread via intravenous injection. These means of transmission are called sexual transmission and blood transmission. Now that we know how these diseases spread from person to person, we can protect ourselves by using condoms, by saying no to drugs, and above all by having sexual relations with only one partner. Thorough testing now ensures that the blood used for transfusions in the United States is safe.

TRANSMISSION BY FOOD AND WATER

Microbes are everywhere, including in food and water. The food and water supplies in the United States are as safe to consume as those anywhere else in the world. Nevertheless, they are not completely safe. Occasionally, infection spreads when people eat food or drink water contaminated by human or animal feces. A widely publicized occurrence of this type of disease transmission occurred in 1992 and 1993: 501 persons had diarrhea after eating hamburgers at restaurants belonging to a northwestern United States fast-food chain. Many victims had bloody diarrhea, forty-five developed kidney failure, and three died. The bacteria that infected these people belonged to a strain of the common intestinal bacterial species *E. coli*, called *E. coli* O157:H7 that colonize the intestines of healthy cattle. When meat from cattle was ground to make hamburger, a tiny portion of intestinal waste apparently found its way into the grinder, which contaminated hamburger sold to the fast-food chain. These bacteria survive cooking if the hamburger is not thoroughly cooked. Once ingested, the microbes survive their passage through the stomach, attach to the large intestine, and grow and make toxins that cause bloody diarrhea and kidney disease.

This disease was known for a decade before 1992, and *E. coli* O157:H7 is probably not new. Why then did this disease emerge with so many cases in 1992? To keep customers happy by providing quick service, the cooks did not cook the hamburgers sufficiently, and the centers were too rare. Thus, it was the change in cooking temperature that permitted the bacteria to survive and infect the customers.

Our water supply can also be a source of pathogenic microbes. Before vaccines eliminated poliomyelitis (polio), a disease of the spinal cord and brain from the Americas, it was a common water-borne infection that entered the water supply via improperly treated human sewage. The polio virus in drinking water infected the cells of the intestines, then later caused the disease. The oral route of infection just described is still common among intestinal viruses, bacteria, and some parasites. This means of transmission is called fecal-oral or food- and water-borne.

MICROBES IN THE MODERN AGE

If you were a microbe, how would you choose to travel? Some infections, such as Lyme disease, that need birds, mice, ticks, and deer to survive are not likely to move very far. But others can take advantage of modern travel and global markets. In the days of sailing ships, microbes traveled as small shipboard epidemics; however, most rarely managed to travel very far. Today, infected people, mosquitoes, mice, and other vectors—and microbes themselves—can travel almost anywhere in the world within a day. Among the diseases that

Water supplies can also be a source of microbial contamination.

can travel this way are HIV/AIDS, influenza, the common cold, and almost any other microbial infection. We should be most concerned about such diseases as new types of influenza and Ebola virus that have the capacity to move, to spread from person to person, and to kill. Fortunately, our public health officials are increasingly vigilant and are likely to find these infections before they cause major epidemics.

How do different microbes travel, and why is it important to understand the different modes of transmission?

Today, people travel around the world all the time. How does this affect the spread of infection, and what additional steps need to be taken to prevent it?

If undercooked, red meat can spread pathogens.

The immune response, like the successful relationship between basic research and applied science, depends on the principals of reciprocity and the effective communication between many participants.

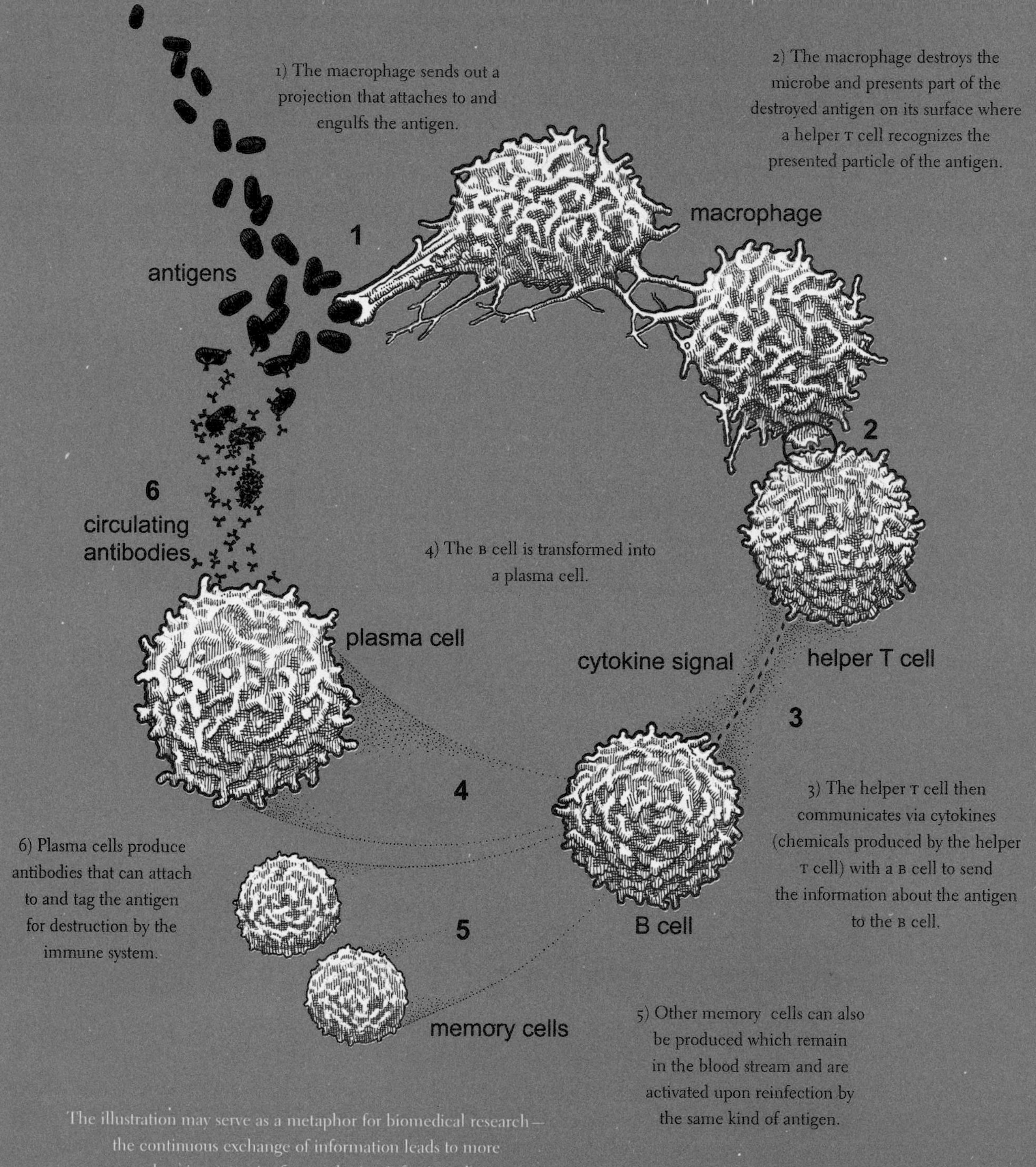

The illustration may serve as a metaphor for biomedical research—the continuous exchange of information leads to more comprehensive strategies for combating infectious disease.

Immunology and Reciprocity

BARRY R. BLOOM

Reciprocity, recognized from the time of Confucius as a fundamental value in human relationships, has not usually been regarded as important to the scientific endeavor. Yet, few areas of biomedical science have provided greater opportunities for reciprocal interaction between science and the real world than has immunology. Basic knowledge of the immune system is clearly essential for shaping new efforts to resist infections, to destroy tumors, and to relieve allergies and autoimmune diseases (diseases resulting from the body's immune response against its own tissue).

In that sense, immunology fits into the generally accepted view that basic science often contributes to applied science, but that the reverse rarely occurs. Crucial support for this general view also comes from the appreciation that many of the most practical applications of immunology—diagnostic tests, vaccines, and therapeutic interventions—derive from fundamental research undertaken with no obvious practical application in mind. To cite just one example, basic studies of how white cells communicate with each other led to the identification of the receptor on T cells that is the site attacked by the AIDS virus. (T cells are the white blood cells responsible for the body's immune response to specific "foreign" molecules.)

Many in the scientific community are concerned because society seems to be placing increasing emphasis on solving real-world problems without recognizing that useful solutions cannot arise simply through the congressional appropriation of funds directed against the problem, but instead depend on basic knowledge not yet acquired. In modern times, that knowledge has derived from the imagination of scientists with the freedom to pursue original ideas independent of their perceived immediate utility. If science appears always to lag our needs, it remains ahead of everything else.

On the other hand, however, much of our understanding in immunology has come from clinical medicine and efforts to master disease. Here, applied science has contributed to basic science. For example, studies of children who have rare

Barry R. Bloom is Dean and Professor of Immunology and Infectious Disease, Harvard School of Public Health.

genetic deficiency diseases that make them highly susceptible to infection helped researchers to explain the crucial role of the thymus (a gland in the neck that helps produce T cells) and antibody-producing cells in immunology.

Through this reciprocity between basic and applied science has come the ultimate application of immunology to improving the human condition—vaccines. Economists know it costs much less to prevent disease than to cure it, and vaccines are one of the most cost-effective interventions to prevent death and disease. In our own country, President Bill Clinton has proposed a comprehensive childhood immunization initiative. The program aims to eliminate an impressive list of diseases by the end of this decade: diphtheria, tetanus, poliomyelitis, measles, rubella (German measles), mumps, whooping cough, and bacterial meningitis. It would reduce pneumonia and influenza in adults older than sixty-five years. It will promote the development of immunological interventions, which are vaccines in the broadest sense, not only against infectious diseases such as AIDS, but also against allergies, arthritis, multiple sclerosis, and cancer.

Beyond the borders of the United States, childhood immunizations have saved more than three million lives. Since 1974, the efforts of the United Nations International Children's Emergency Fund (UNICEF) and the World Health Organization increased the number of the world's children receiving immunizations by the age of one from fifteen to eighty percent. It is shocking that in 1982 only about half our children received their recommended immunizations by the age of two. In such cities as Houston and Miami that figure was less than thirty percent. At another level for reciprocity, then, we have much to learn from developing countries that protect their children.

In his book *Preparing for the Twenty-First Century*, Paul Kennedy identifies the major issues to be addressed as we head into the next century as equity, population, and the environment. In a curious way, immunology may have something to contribute to each. There are enormous disparities in quality of life of people born in different parts of the world. The Third World contains three fourths of the Earth's population. There, eighty-seven percent of all births and ninety-eight percent of all infant and childhood deaths occur. One in ten people suffers from a tropical disease; 190 million children are undernourished; and ten million people die of acute respiratory and diarrheal infections each year.

At the same time, the United States gets many of its raw materials from the Third World. In fact, our trade with developing nations exceeds that with Western Europe and Japan combined and is our most rapid area of growth. In recognition of both the needs and the opportunities, seventy-one heads of state, an unprecedented number, met at the United Nations in 1990 for the World Summit for Children. They pledged to put the health and education of children at the top of the international agenda. Immunology and vaccines have a special role to play. By the end of the decade, paralytic poliomyelitis can be eliminated and measles can be controlled. Although pathogenic protozoa are the largest single source of infection, no vaccine against a protozoan that infects humans currently exists. Clearly, the need is urgent for vaccines against malaria, leishmaniasis, and other diseases caused by protozoa.

The argument has often been heard that keeping children alive in the Third World merely increases suffering from poverty and malnutrition. I would point out that in no country have birth rates declined before a decline in death rates and that one of the most powerful forces for reduction in fertility is child survival; keeping children alive so that parents know they will be taken care of in old age.

In reflecting on some reciprocities between immunology and the real world, I am ineluctably drawn to the conclusion that biomedical research, if allowed to flourish, has incalculable potential for humane contributions, and to Oscar Wilde's view that "a map of the world without Utopia on it is not worth glancing at."

Why is immunology an opportunity for reciprocal interaction between science and the real world?

How are the use of vaccines in the Third World an example of how biomedical research can increase the potential for reciprocity and improve world health?

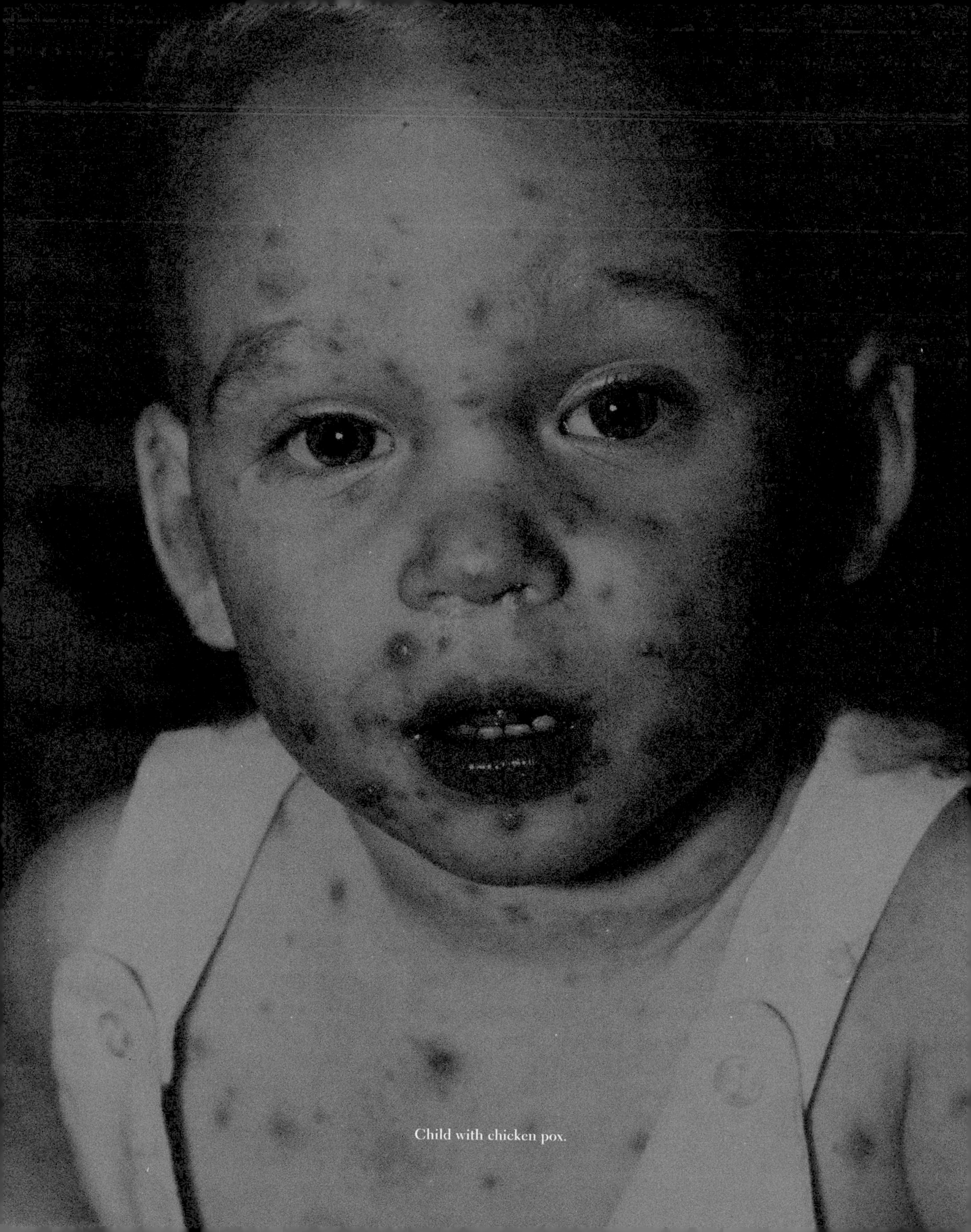

Child with chicken pox.

Studying Infection in the Clinic

ANNE A. GERSHON

One of the most impressive advances in twentieth-century medicine has been the conquest of infectious childhood diseases through immunization. This conquest required both the laboratory development of vaccines and the testing of these vaccines by dedicated clinician scientists. These clinicians vaccinated children in their care and then monitored their health to see if the new vaccine gave them immunity to a specific disease. In this way, the medical community developed vaccines against poliomyelitis, measles, mumps, German measles, meningitis, whooping cough, diphtheria, and tetanus. Today, the United States is virtually free of these diseases, although on an international level they still are not conquered. In August 1998, an article in the *New York Times* questioned whether today's medical students need to learn about most of these diseases—so rare have they become! In the future, recently developed vaccines against chicken pox, diarrhea, and influenza are likely to have a significant impact on disease. Currently, however, these vaccines are so new that knowledge of their influence is still incomplete.

The testing of most of the vaccines mentioned occurred in clinical trials conducted in physicians' offices. Because many children visit pediatric clinics, these clinics are ideal for studying the safety and effectiveness of vaccines in both normal and "immune impaired" children. Children with chronic illnesses that predispose them to severe infections are also frequently seen in the clinic so that their progress after vaccination may be followed closely. Today we refer to such vaccine studies, which go back about a 100 years, as clinical trials. Health experts use modern clinical trials to study the safety and effectiveness of a potentially useful vaccine.

According to the usual testing methodology, half the patients in the test group receive the vaccine, and the other half, who are called the control group, receive a placebo. A placebo is an inactive substance that does not fight infection, but that appears identical to the real vaccine. To avoid observational

Anne A. Gershon is Director of the Division of Pediatric Infectious Disease at Babies and Childrens' Hospital in New York and Professor of Pediatrics at Columbia University College of Physicians and Surgeons.

bias when testing a vaccine, the nurses and doctors involved with the patients, and the patients themselves, do not know whether the substance injected is the vaccine or placebo. After injecting the vaccine or placebo, the nurses or doctors count the number of children who become infected in each of the two groups. A vaccine is successful if it satisfies two conditions: (1) none, or at most a few, of the vaccinated people develop the disease symptoms, and (2) a statistically significant number of people who received the placebo do develop the illness. To conduct such clinical trials on children, the health care provider is required to obtain "informed consent" from the parents of the test subjects. The tester must clearly describe the test procedures and the risks and possible benefits of participating in the clinical trial. The parent and child can then decide rationally whether to participate in the study. Participation in such studies is purely voluntary. Patients cannot be forced to participate against their will.

In 1995, the U.S. Food and Drug Administration (FDA) licensed a vaccine against chicken pox (varicella) for routine use in the United States. The following is a brief account of the development of the vaccine and its licensing by the FDA. Because this vaccine was tested in various physicians' offices—in the process described above—this account serves as an example of scientific advances made through medical research conducted in clinics.

In the United States today, an estimated four million cases of chicken pox occur annually, roughly the same number as in the prevaccine era. Most of these cases occur in children younger than ten. The pathogen that causes chicken pox is the varicella-zoster virus (VZV). The symptoms of chicken pox, which typically last about one week, are fever, irritability, fatigue, loss of appetite, and an extremely itchy, blistery skin rash all over the body. The most common but definitely not the most serious complication of chicken pox is an additional infection by bacteria; this happens in about one of five hundred cases. In an estimated one in four thousand cases, a more severe complication occurs: mild to severe encephalitis (inflammation of the brain). About 9,000 children and adults infected by VZV become hospitalized annually with chicken pox, and about 100 of these patients die of the disease. Most of these individuals were healthy before getting chicken pox. Children with impaired immune systems, such as those who have various forms of cancer or have undergone organ transplants, are at risk to develop severe chicken pox. Finally, about two percent of the offspring of mothers infected by VZV during the first or early second trimester of pregnancy have rare types of birth defects caused by the virus.

VZV is an unusual virus because once it infects a host, it can cause the host to have two completely different diseases that typically occur years apart: chicken pox and the disease that physicians call zoster or herpes zoster, commonly called shingles. Shingles is an illness in which patients develop a blistery painful rash confined to one side of the body. It occurs particularly in the elderly and in individuals with chronic illnesses that impair the functioning of their immune system. Shingles is due to the reemergence of VZV that entered the nervous system during chicken pox and then "hid" in the body in an inactive or latent state. Reactivation of VZV from its inactive state occurs when cellular immunity to VZV wanes. This commonly happens in people over the age of sixty. About fifteen percent of the population develop shingles over their lifetime; about 500,000 cases occur annually in the United States. "Post-herpetic pain" develops in the area of the shingles' rash after the rash itself has healed and is a common complication of shingles in the elderly. This may lead to a prolonged period of misery and debilitation.

During an infection by VZV, the body develops two forms of immunity antibodies and cell-mediated immunity that help destroy the infecting VZV particles and prevent reinfection by the same virus. Cell-mediated immunity, also called cellular immunity or T cell-mediated immunity, involves processes that destroy viral particles or other parasites "hidden" in cells by destroying the cells containing them. About one to two percent of U.S. adults lack immunity to VZV because they did not have chicken pox as children. To prevent the potentially serious consequence of VZV infection, these adults should receive a vaccination against the virus. An infection by VZV is about twenty times more likely to cause deaths in adults than in children.

The first steps toward the development of a vaccine against VZV probably date from the early twentieth century. During that period, some physicians inoculated children with blister fluid from patients with shingles. The aim was to produce a mild case of chicken pox in the child. If all went well, the child would develop an immunity that would protect him or her against future natural exposures to the virus. Experimental inoculation attempts usually led to mild infections, which suggested that developing a vaccine against VZV was feasible. Today, the medical community rejects such studies as these, and they claim them to be "unsafe" or "unethical."

Vaccination against chicken pox became possible only after researchers learned how to grow the virus artificially in the laboratory. (In the middle of this century, Nobel Laureate Dr. Thomas H. Weller and his colleagues were the first to establish VZV as the cause of chicken pox. They isolated VZV from cultures of human cells. Weller was not trying to produce a vaccine, but merely to grow the virus artificially.) Dr. Michiaki Takahashi and his colleagues developed the first successful vaccine against VZV in the early 1970s, using cell culture methods similar to Weller's. This vaccine is similar in concept to the Sabin oral polio vaccine and to the vaccines against measles, mumps, and rubella (German measles), because it is a "live attenuated" vaccine. That means it contains a virus that is "attenuated" (made into a less virulent strain) in the laboratory. These vaccines either do not cause disease at all or—in a small minority of vaccinated people—cause a very mild, and harmless, form of the disease. Nevertheless, they induce the formation of protective antibodies against the natural (and virulent) strains of these viruses; they also stimulate cell-mediated immunity.

Testing of the live attenuated VZV vaccine began in the United States in the late 1970s, driven largely by concern about chicken pox's severe effects on leukemic children. About ten percent of such children who developed chicken pox died of it. (Clinical pediatricians from the United States and Canada led the Varicella Vaccine Collaborative Study Group that carried out these studies.) The National Institute of Allergy and Infectious Diseases (NIAID) provided financial support for this group. Eventually, the Collaborative Study Group immunized more than 500 children with leukemia in remission. Studies showed that the side effects of vaccination were rarely severe, and always treatable. Thus, the vaccine was considered "safe." The VZV vaccine provides protection not only against chicken pox but against shingles as well.

To learn if the vaccine was effective, it was necessary to measure antibody levels against VZV, and the degree of cellular immunity to it. The VZV vaccine was very successful in preventing chicken pox from developing after exposure to VZV. No immunized leukemic children died of chicken pox; most did not develop chicken pox, even after exposure to an unvaccinated brother or sister with the illness. The injected vaccine stimulated the immune system to develop antibodies and cellular immunity to VZV without causing any illness. These antibodies and the resultant cellular immunity then prevented chicken pox from occurring after exposure to VZV. Good vaccines not only stimulate the immune system immediately, they also cause a long-lasting immune response that results from the development of "memory lymphocyte." These cells "remember" that they have seen a virus already and then they recruit the rest of the immune system into action if the virus tries to invade the body a second time.

Because the VZV vaccine appeared to work in children with leukemia, healthy children received the vaccine in clinical trials in the mid-1980s. These trials too occurred mainly in physicians'

Dr. Anne Gershon immunizing her colleague Dr. Philip La Russa, now Associate Professor of Clinical Pediatrics at Columbia University. He was the first American adult ever to receive the chicken pox vaccine.

offices and included close to 10,000 children and adolescents. One double-blind placebo-controlled study, involving only a small number of children, showed the vaccine to be almost 100 percent effective in preventing chicken pox. The sponsor of these studies was the manufacturer of the vaccine, Merck and Co. The vaccine proved to be extremely safe. The main side effect was temporary discomfort at the site of the injection. About four percent of those vaccinated had a mild, generalized rash with an average of five skin bumps or blisters. The vaccine was highly immunogenic. After one dose of vaccine, almost 100 percent of the children developed antibodies and cellular immunity. However, additional studies other than the one already mentioned showed that the vaccine did not provide full immunity for all its recipients. It provided complete immunity to eighty-five to ninety percent of the children receiving it, and the remaining ten to fifteen percent developed extremely mild cases with no complications. Follow-up studies have monitored some of these test subjects for up to ten years after their vaccination; most of these still have immunity to VZV. The monitoring of some people vaccinated as children or young adults will continue for many years to learn how long the immunity provided by the vaccine lasts.

The results of this vaccination study led the American Academy of Pediatrics and the Advisory Committee on Immunization Practices (ACIP) of the Centers for Disease Control and Prevention to recommend the following: routine immunization of healthy children older than one year, of healthy adolescents and adults lacking immunity to VZV. Children younger than thirteen should receive one dose of vaccine; individuals who have had their thirteenth birthday should receive two doses of vaccine. Although the FDA licensed the vaccine for use in the United States in 1995, and its use is increasing, it is still not widely used. Health officials hope that eventual widespread use of the vaccine will cause the number of illnesses and deaths from VZV to approach zero.

This brief account of the development and application of a successful live attenuated VZV vaccine reveals a complex and demanding process involving both laboratory scientists and clinical health care workers. Laboratory scientists started the process by isolating and cultivating the virus and by developing a live attenuated virus. A drug company developed procedures to produce the vaccine in quantity. Clinical practitioners completed the process by successfully testing the vaccine and by subsequently routinely administering it to all Americans susceptible to chicken pox. The story of this vaccine's journey from lab to market is only one example of the crucial role the nation's health care clinics play in understanding illness and helping to prevent it.

What is the process used in modern clinical trials that study the safety and effectiveness of a potentially useful vaccine?

How is the testing of the chicken pox vaccine an excellent example of the important role that clinics play in the development of vaccines?

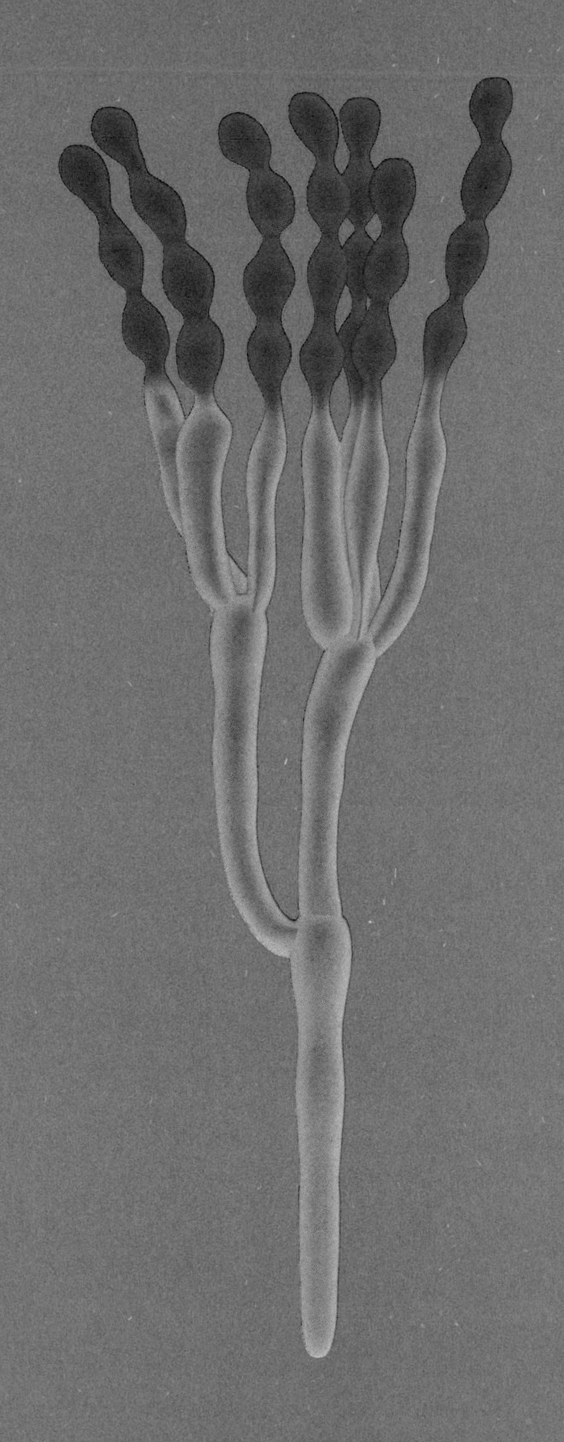

Computer-generated image of the mold, *Penicillium notatum*.

Discovery Scientists: Developing Drugs to Fight Disease

RICHARD J. COLONNO

INTRODUCTION

The universe of disease-causing viruses, bacteria, and fungi is growing. Worse, many pathogens we have held in check are beginning to resurge as they become resistant to current drugs. (Because of the uncanny ability of microbes to mutate, infectious diseases are the only diseases for which potent drugs become obsolete.) Winning the battle against these pathogens requires the development of new drugs that inhibit different microbial processes or different stages of their replication. Discovery scientists are the researchers who discover and develop drugs to fight the invisible enemies we cannot see. It is a challenging mission that easily discourages those who do not have the perseverance and commitment to overcome the many disappointments and setbacks they will encounter along the way. However, the discovery of a truly novel inhibitor exhilarates the discovery scientists with an excitement that few ever experience.

BASIC STRATEGIES FOR ANTIMICROBIAL DRUG DISCOVERY

To discover a new drug effective against a specific microbe, discovery scientists must first isolate and/or identify the enemy microbe, discover the class of microbes to which it belongs, and learn how it acts in the host it infects. Many pathogens of human disease have been known for years; others, such as the AIDS-causing pathogen HIV, have emerged only recently.

Next, discovery scientists must devise a strategy for discovering an inhibitor that will selectively attack the microbe without hurting normal human

Richard J. Colonno is Vice President, Infectious Diseases Drug Discovery, Bristol-Myers Squibb Company, Wallingford, Connecticut.

cells. Viruses can exist only as parasites because they can replicate by using a host's cellular machinery. Therefore, they present a different challenge than do bacteria and fungi, which are true organisms that can self-replicate. The strategy chosen by the researchers typically falls into either of two general categories: the traditional whole-cell approach or the biochemical approach.

The traditional whole-cell approach establishes the rate at which a microbial population grows in a particular medium—on a petri plate or in a cell culture system—and then searches for substances that inhibit this growth; this type of test is a "growth or cell protection assay." We call this approach the whole-cell approach because it tests potential inhibitors using intact (or "whole") target microbial cells or cell culture systems. The whole-cell approach is logical for those organisms that grow readily under laboratory conditions and that are not so dangerous that researchers need to take elaborate precautions (e.g., sealed-off chambers, and the like) to prevent them from escaping into the surrounding environment. The advantage of this approach is that discovery scientists gain direct evidence that a particular substance inhibits the targeted pathogen, at least under laboratory conditions; such evidence is often a major milestone along the pathway to drug discovery. Many of today's antibiotics—including penicillin, streptomycin, erythromycin, and tetracycline—were discovered decades ago using this traditional approach.

The alternative to the traditional whole-cell approach is the biochemical approach, which discovery scientists use when growing the targeted organism in the laboratory is difficult. The key to the biochemical approach is to identify a component of the pathogen, usually a protein, that is essential for the pathogen's growth and replication but that differs chemically from related proteins in human cells. (Establishing that a specific protein is essential involves genetically altering the specific gene encoding the protein to disable or delete it; if the genetically altered pathogen can no longer grow, its growth depended on that protein.) This essential protein is then a good target for intervention, because a substance that blocks its ability to function, will normally prevent the growth of the pathogen containing it.

To inhibit the targeted protein, discovery scientists need to learn its function and develop a test that they can use to search for inhibitors. For example, the targeted protein might be a protease (pronounced "pro-tee-ace"); a protease is an enzyme, and therefore a protein itself, which has the function of cleaving other proteins into smaller fragments. Here, investigators would develop a functional assay in which the protease itself is able to cleave other peptides or proteins; they would then look for substances that inhibit its protein-cleaving action.

SCREENING FOR INHIBITORS

Whether discovery scientists are targeting a specific protein of a pathogen (the biochemical approach) or the entire pathogen (the traditional whole-cell approach), searching for a novel growth inhibitor is the equivalent of "looking for a needle in a haystack." Therefore, they must convert any test for inhibitors to one that can test hundreds of thousands of potential inhibitors quickly. Through a combination of automation, robotics, and new highly sensitive detection technologies, they can convert a test with capacity limited to only several samples a day into a screening technique that can handle thousands of test samples a day. Sources of potential inhibitors are many, but most inhibitor molecules usually fall into two major categories: chemical molecules (synthetic chemicals) accumulated from programs mostly unrelated to infectious diseases, and biological molecules, such as substances produced by microbial fermentations or extracted from plants, herbs, sponges, and even snake venom. In the search for new drugs, few sources of starting material escape our attention; the best drugs of tomorrow may be substances totally unknown today.

As described up to this point, the strategy for drug discovery might seem quite simple—identify the organism or protein we want to inhibit, and then use automated techniques to screen many thousands of molecules to find one that possesses inhibitory activity. In reality, these are only the first steps. The discovered inhibitor, usually called the primary screen hit, must now be put through a battery of secondary tests to ensure that it is a selective inhibitor of the targeted protein or pathogen. At this stage of the discovery process, many substances that acted as inhibitors in the screening test fall by the wayside. Primary screen hits, the discovered inhibitors, receive a thorough evaluation at this stage to ensure that their inhibitory effect results from a highly specific biochemical mechanism. In traditional screening (the screening of substances acting on intact cells), that means ensuring that the inhibition is not due to the non-selective killing of all types of cells; it is also necessary to learn the precise step in the cell's life cycle that is inhibited. For biochemical screen hits, this secondary evaluation must prove that the inhibitor is truly selective for only the targeted protein, and that it works against the intact pathogen in a traditional type of growth assay. All too frequently, the results of this labor-intensive screening process yield few useful inhibitors and require a return to primary screening, perhaps involving the selection of an alternative target protein. For the precious few hits that survive scrutiny by secondary evaluation, the challenge now becomes the discovery of a true "drug lead" among these early screen hits—through the procedures described presently.

BYPASSING SCREENING

Rational drug design is an alternative approach that sometimes allows scientists to bypass the screening process. Here, discovery scientists must first determine the molecular structure of the targeted protein; that is, the three-dimensional arrangement of its component atoms. Knowing the exact three-dimensional structure of the target protein, discovery scientists could, in theory, "design" a molecule that would have just the right shape so that all or part of it could lodge in a small "crevice" of the target protein and inactivate it in the process. In practice, instead of designing new molecules, they typically use high-powered computers and sophisticated software programs to search virtual "chemical libraries" for already existing molecules that have the desired geometry and other properties to inhibit the target protein. These "designed" inhibitors are then obtained (usually through chemical synthesis) and tested in the functional tests described above. If they show an ability to inhibit the growth of the organism containing the target protein, researchers put them through the same series of secondary tests used for substances found as screen hits—to ensure that they are truly working through the biochemical mechanism predicted in their design.

Another way of bypassing an empirical screening approach is to target a class of proteins for which known inhibitors already exist. Again, proteases can serve as a good example. We know a great deal about how proteases cleave other proteins, and we already have a selection of known protease inhibitors; therefore, we can attempt to modify and customize these existing protease inhibitors to generate new inhibitors that are selective for a particular pathogen's protease while not inhibiting the human body's normal cellular proteases.

FROM SCREEN HIT TO CLINICAL TRIAL

Progressing from a screen hit to a "drug lead" (as in a "lead" that helps a detective solve a crime) involves the close collaboration among a team of biologists and chemists. Typically, an inhibitor molecule that survives secondary evaluation must be "engineered" if it is to become an effective drug. Again, this requires a strategy. Discovery scientists establish objectives covering potency (how powerfully it acts against the target pathogen), eventual formulation (whether physicians will administer it orally or through intravenous injection), spectrum (the range of microbes it inhibits), solubility, stability, lack of toxicity, and other pharmacological parameters for the ideal drug. Potency is usually the initial primary focus, because a very potent drug can sometimes overcome deficiencies in the other areas, and will usually delay the emergence of resistant variants. Hundreds, if not thousands, of related molecules must be prepared and tested to find candidates that satisfy the objectives. Along the way, discovery scientists test many of these compounds in appropriate animal models, if available, to learn if the drugs can protect these animals from infection by the targeted pathogen. This empirical process of optimization and further modification continues until investigators have found a particular series of drugs that satisfy most, if not all, of the criteria set (a "lead series"). The final step is selection of a "candidate drug" for clinical testing on humans. Large quantities of the new compound are subsequently prepared using protocols approved by the U.S. Food and Drug Administration, and tested for safety. Upon successful completion of these tests, physicians will test the compound in humans in a series of clinical trials that will decide whether it is safe and effective.

CONCLUSION

Although the odds against finding new microbial inhibitors remain long, they have recently become shorter; our greater understanding of these pathogens' life cycles and genome sequences, coupled with our application of new technologies, has provided us an impressive list of successes. Powerful drugs such as those discovered and developed over the past few years to treat HIV infection serve as a prime example; these drugs have greatly lengthened the average survival time of those infected. These successes and a steady stream of others encourage discovery scientists to press forward in their quest for miracle drugs against a variety of pathogens.

Why is it necessary to take such stringent steps in the testing of drugs before a drug is allowed to be used against an infectious disease?

Why is the search for new microbial inhibitors such an elusive goal, even in this generation of biomedical engineering and drug discovery?

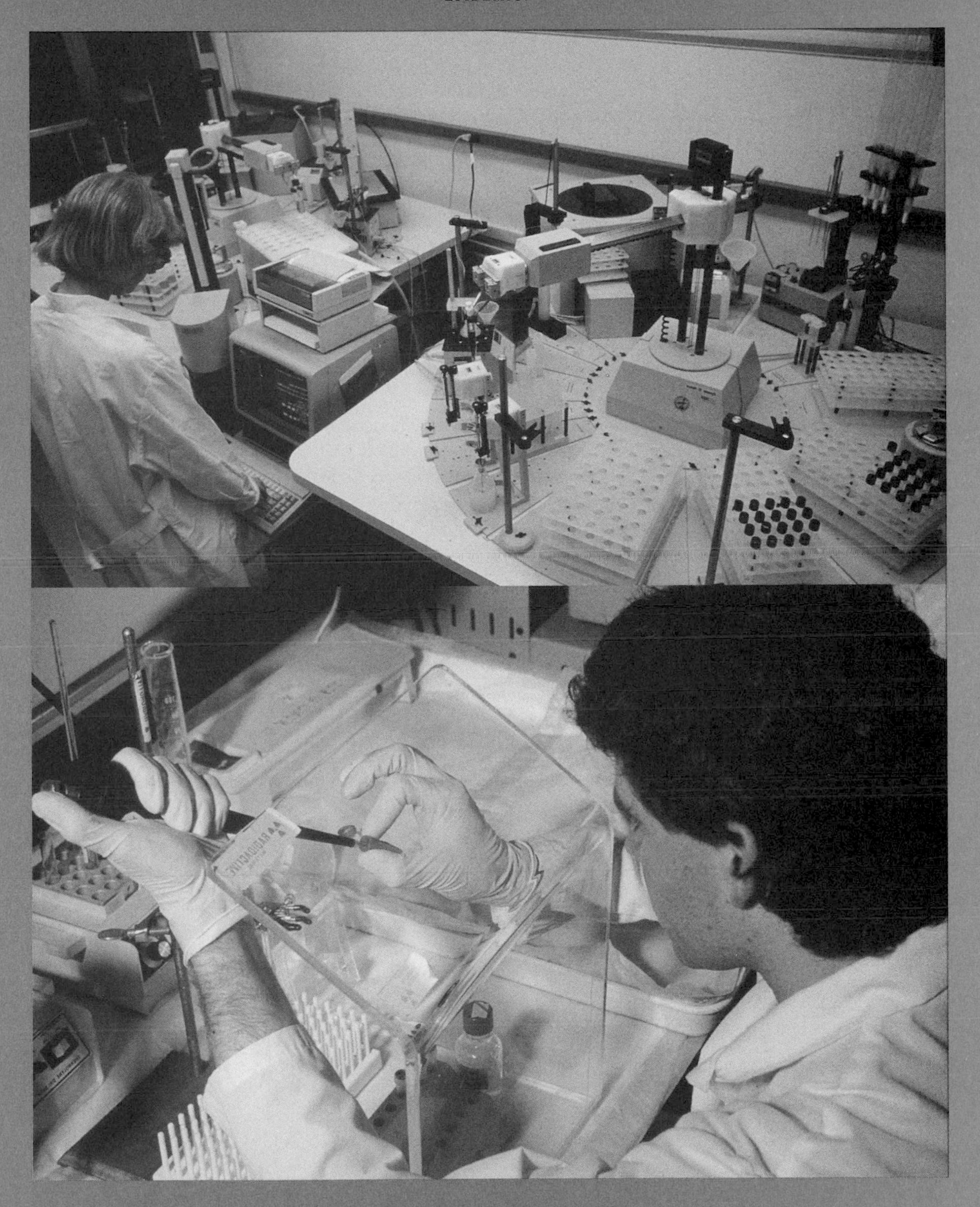

Discovery scientists at work in modern laboratories.

Have you gotten your flu shot yet?

In autumn, you should get one if you are over sixty-five; if you have diabetes; if you have heart, lung, or kidney disease; if you have an immune system disorder; if you are a woman who will be in your second and third trimester during flu season; or if you live with or take care of anybody in any of these high-risk groups. Students living in a dormitory or members of the armed services living in a barracks should also be vaccinated to prevent disruption of routine activity during outbreak. Even if you do not fall into any of these risk categories, you may wish to get one anyway, unless you are allergic to eggs.

Flu is short for an illness caused by an influenza virus. Although the common cold is also caused by a virus, it's not an influenza virus and the flu is more than just a very bad cold. The usual symptoms are high fever, chills, weakness and extreme fatigue, painful head and muscle aches, cough and sore throat, and loss of appetite. It is transmitted from person to person through particles created by sneezing, coughing, and talking. For most people, a bout of flu means a week or ten days of feeling really lousy, but between 20,000 to 40,000 people die of the flu each year in the United States; nine out of ten of them are over sixty-five years of age.

If the flu is caused by a virus like chicken pox or measles, why doesn't getting it once or being vaccinated make a person immune for life? The answer to that question also explains why the Influenza Branch of the Centers for Disease Control and Prevention (CDC) in Atlanta, Georgia, is a beehive of activity twelve months of the year.

We call it the flu, but it would be more accurate to say flus. Viruses are masters of mutation, and the influenza virus is one of the most changeable. No sooner has an immune system manufactured antibodies against a particular strain than the next generation of virus changes just enough to evade recognition by the antibody. Formulating a vaccine that will be effective against the newest version of the flu involves educated guesswork based on data gathered by a worldwide network of flu watchers.

The World Health Organization (WHO) oversees global influenza surveillance and the strain selection process that determines the ingredients of each year's flu vaccine. This system, which has been in place since 1947, relies on 108 national reporting centers in seventy-three countries where influenza viruses are collected and analyzed. The national centers send viral samples and their preliminary analysis to one of four international reference centers: the Commonwealth Serum Laboratories in Melbourne, Australia; the National Institute for Medical Research, in London, England; the CDC; and the National Institute of Infectious Diseases in Tokyo, Japan. There each virus is subjected to more detailed examination of its antigenic properties. The aim is to characterize each one and compare it to those that have been around in past years to see whether, and in what ways, it has undergone mutation. Assays in which the virus is exposed to antibodies developed by people who had the flu vaccine during the fall provide further hints about how much a particular viral strain has changed.

Each February, WHO holds a meeting at which representatives from the four reference centers share their data. Out of that meeting comes a recommendation for what the components of that year's flu vaccine should be. In recent years, the vaccine has been what is termed trivalent—containing three viral strains, which have been rendered incapable of causing infection but still able to stimulate the production of specific antibodies. Usually, only one or at most two components are changed each year.

Health authorities in each country are free to decide whether to follow WHO's recommendations. In the United States, the decision is made by a Food and Drug Administration (FDA) advisory committee, which has historically gone along with WHO. The FDA provides pharmaceutical manufacturers with the selected viral strains, and a bizarre production process begins.

It turns out that the best place to grow a virus is inside a fertilized chicken egg. A small amount of virus is injected through the shell of the egg, where it rapidly replicates, producing millions of copies of itself.

Multiply this by millions of eggs, and you have a harvest of about eighty million doses of flu vaccine, just in time for flu season, which begins in December and peaks during January and February. Manufacturers aim to have vaccine available by mid-September, to give people a chance to get their shots before the first cases of flu appear. (Now, you know why people with egg allergies shouldn't get flu shots. If even a tiny bit of egg protein finds its way into the vaccine, it could cause an allergic reaction more dangerous than the flu.)

How good is WHO at predicting the new flu? Pretty good, but not perfect.

Flu watchers keep an eagle eye on China, which is where most new viruses first appear. No one is sure why, but it's probably a combination of the huge numbers of people living very close together and the ease with which influenza viruses exchange genetic material.

In addition to humans, swine and fowl also get the flu. If a pig flu virus comes in contact with a human flu virus, they may trade genes in a process called reassortment. One or both of the new viral forms might then make a trade with a nearby chicken virus. Before long, a completely new virus capable of infecting chickens, pigs, and humans is making the rounds, and none of the animals has antibodies to combat it.

When a new virus is reported in China, epidemiologists watch to see if it is spreading. They may have as long as twelve to eighteen months advance warning, or the virus may not spread at all. But sometimes there are nasty surprises. During the 1997–98 flu season, for example, the predominant flu strain to hit the United States came not from China, but from Australia. Because the Australian strain was identified in the summer of 1997, there was insufficient lead time to plan for its inclusion in the vaccine. Consequently, people who received the flu shot and were infected by the Australian strain may have still become ill.

The components of the 1998–99 flu vaccine were two viruses found in Beijing, China, in 1995 and 1993, and one virus from Sydney, Australia, in 1997.

What's the recipe for next year's flu vaccine? WHO knows.

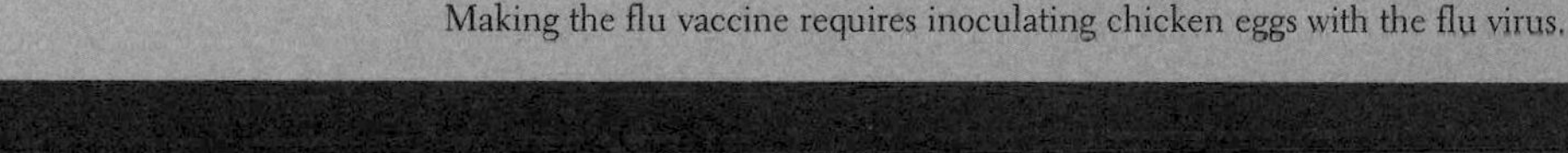

Making the flu vaccine requires inoculating chicken eggs with the flu virus.

Profile

Stuart Levy: The Battle Against Drug Resistance

Stuart Levy, M.D., is a man with a mission. He wants doctors to wash their hands more and prescribe antibiotics less. And he wants the public to understand that drug resistance is everyone's business and solving the crisis it represents is everyone's responsibility.

As President of the Alliance for the Prudent Use of Antibiotics (APUA), Dr. Levy heads an international organization with members in ninety countries dedicated to curbing antibiotic resistance through research and the education of health care providers and consumers. He believes this must be a worldwide effort because antibiotic resistance is a global problem.

"Whether it is in animals, in patients in the hospital, or people at home, antibiotics select resistant bacteria that enter a common environmental pool," he explains. "Antibiotic use causes ecological changes both in the individual (man, animal, plant) and the environment." In an age of international commerce and travel, the environmental pool stretches across the globe.

After graduating from Williams College, Dr. Levy studied medicine at the University of Pennsylvania. Working on a fellowship at the National Institutes of Health, his focus was on bacterial cell wall synthesis. But when he became fascinated by what he calls bacterial ingenuity in outwitting antibiotics, he found his life's work.

Levy moved to Tufts University School of Medicine, where he is now Professor of medicine and molecular biology and microbiology, and Director of the Center for Adaptation Genetics and Drug Resistance. In 1979, he uncovered how certain bacteria resist the widely used antibiotic tetracycline. They are able to pump the drug out of cells so it never reaches its target. Called the drug efflux mechanism, it is now recognized as the basis of bacterial resistance to many drugs.

In 1992, Dr. Levy published a book entitled *The Antibiotic Paradox: How Miracle Drugs Are Destroying the Miracle*, which sounded the alarm about the drug resistance crisis and outlined a battle plan to preserve medicine's ability to fight infection in the face of multi-drug-resistant bacterial strains.

Antibiotics have been hailed as miracle drugs since their introduction in the 1940s. Beginning with penicillin, these antibacterial medicines have saved untold numbers of lives from diseases that had previously been cripplers and killers, including tuberculosis, staphylococcus and streptococcus infections, sexually transmitted diseases such as syphilis and gonorrhea, pneumonia, and meningitis, which affects the brain and spinal cord. They have been used, appropriately, to treat infected wounds, and increasingly for sore throats, ear and urinary tract infections, acne, and many other relatively mild conditions. In the half century of their use, they have often been prescribed to prevent secondary bacterial infections for people with viral diseases, a practice that has led to much misunderstanding about their proper use.

Antibiotics, which act against bacteria, have no power over viruses, be they HIV or the common cold. Nonetheless, many people expect their doctors to prescribe medicine for whatever ails them. Overuse and misuse of antibiotics is a major factor in the evolution of new strains of drug-resistant bacteria.

The paradox, as Dr. Levy sees it, is that these powerful drugs have given rise to more powerful bacteria, the so-called superbugs that are invulnerable to many of the more than 100 antibiotics now on market. We are running out of drugs to treat such life-threatening diseases as TB, meningitis, and staphylococcal septicemia, and are seeing more frequent outbreaks of food-borne bacteria such as salmonella and *E. coli*, often with deadly results, as well as difficult-to-treat community-acquired infections such as ear infections and meningitis with the pneumocacas and sexually transmitted gonorrhea.

Perhaps the most alarming thing about antibiotic-resistant bacteria is that our hospitals are a major storehouse of the superbugs. Hospital-acquired bacterial infection is a major—and often deadly—complication of medical treatment throughout the world, including in the United States.

The way resistance develops is complex, but in simplified terms it's the old story of survival of the fittest, taken down to the cellular level. When an antibiotic encounters a colony of bacteria, it kills cells that are susceptible to its action. Others survive either because they already have some resistance, possibly acquired through mutation, or because too little of the antibiotic drug was given to wipe out the entire colony. (That might be the case if the patient stops taking the prescribed medication because he or she "feels" better and wants to keep the rest for the next time a "bug" strikes.) The surviving bacteria proliferate, and they do so more freely because there is less competition from cells that are susceptible to the antibiotic. Within a few bacterial generations—a matter of days, not years—drug-resistant strains make up most of the cell population. And they spread from person to person with every cough, sneeze, and handshake.

What can we do to stem the tide of drug resistance? Here are some of Dr. Levy's suggestions.

- Do not demand antibiotics, even when you feel very sick. Unless a doctor confirms that what's ailing you is a bacterial infection serious enough to require treatment with antibiotics, do not expect to get a prescription.
- If your doctor does prescribe antibiotics, take all the pills. Don't stop taking them when you begin to feel better and do not, under any circumstances, hoard some for another time you feel unwell and think antibiotics might help.
- Practice prevention so that we can rely less on treatment.
- Wash your hands before and after handling food, and more often if you are ill or are caring for someone who is.
- Use warm water and soap, but not antibacterial soaps and washes, unless you have an impaired immune system or are caring for someone who does. Antibacterial personal and household products, which have recently become popular, contribute to the triumph of resistant bacterial strains. Rather than being a boon, they alter the environment in a way that is favorable to the development of resistance—without a doctor's prescription!
- Wash all fruits and vegetables before eating and avoid raw and undercooked animal products, such as eggs, meat, poultry, and fish.
- For doctors and other health care personnel, thorough washing of hands between patients is essential to prevent the spread of bacteria from infected patients to healthy ones.

Dr. Stuart Levy,
Director for Center for Adaptation Genetics and Drug Resistance,
Professor of Molecular Biology and Microbiology, and
Professor of Medicine,
Tufts University.

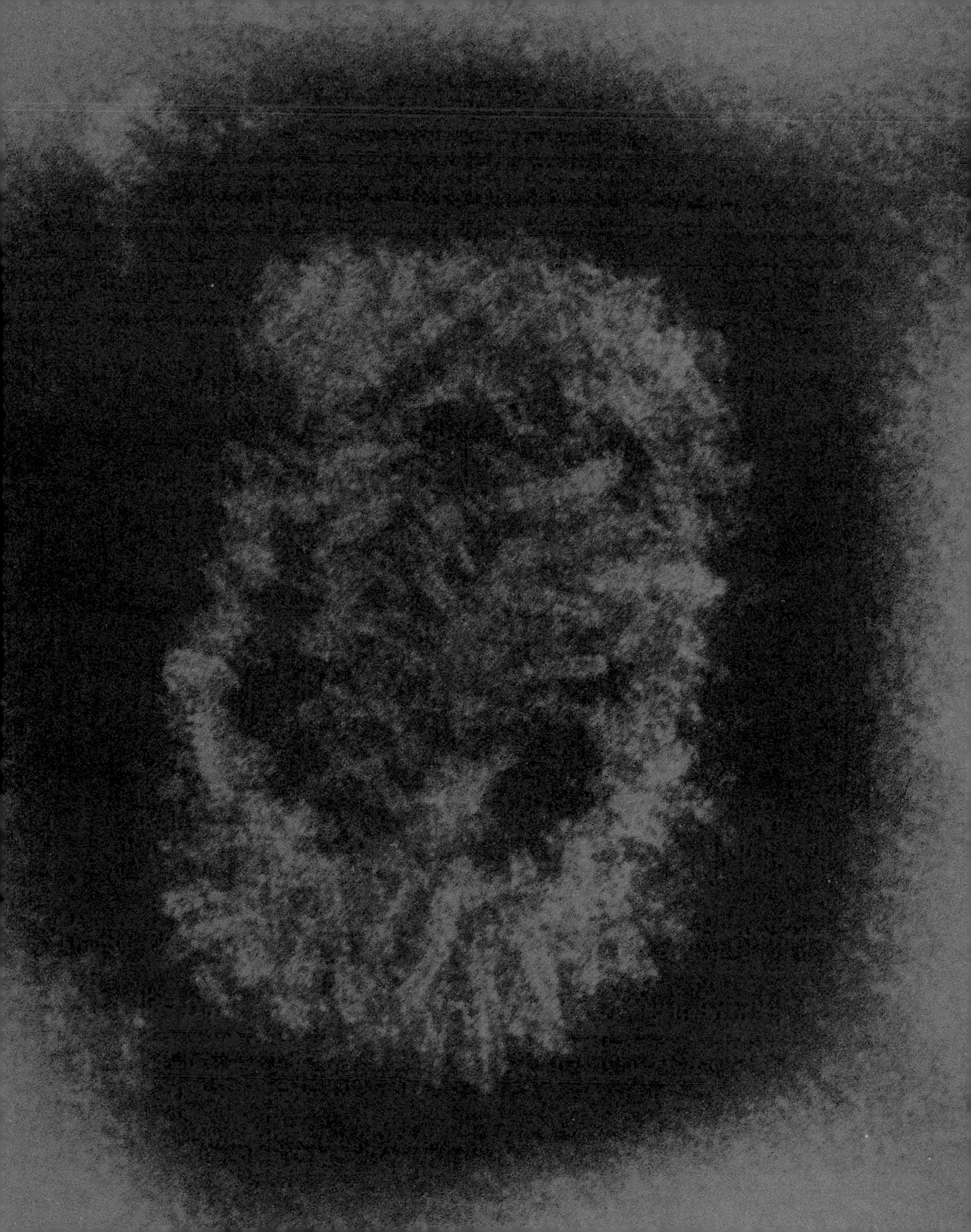

Section Four: Outbreaks

left Electron micrograph of pox (*Poxviridae*).

Workers in Biosafety Level 4 suits.

Section Four: Outbreaks
Introduction

ROB DESALLE AND MARLA JO BRICKMAN

In the preceding section of this book, we discussed the influence of microbes on the human body. In this section, we widen the concentric circles of our understanding of infectious disease to include the spread of infection throughout a population of humans living in a specific region. This is the so-called outbreak stage of an infectious disease. An outbreak is a suddenly occurring and explosively rapid increase in the incidence of an infectious disease in a population. All the cases in an outbreak involve infections from the same source. The major goal of this chapter is to describe the dynamics of an outbreak—the various ways in which an infectious disease spreads throughout a particular population.

Obviously, for an infectious disease to spread throughout a population, the microbe causing it must leave an infected person's body and get into the bodies of others. There are four ways that microbes can leave the infected body. The first is via the respiratory route—through the nose and/or mouth. Most often, infectious microbes will exit this way during sneezing or coughing. These microbes can also leave the body in sputum, saliva, and other mucus-like fluids exiting through the mouth, or in mucus (snot) from the nose. The second exit route for infectious microbes is through blood. The most common way that microbes leave via this route is through bites from vectors such as mosquitoes, or from the use of hypodermic needles. Microbes can also exit in blood from cuts, however microbes exiting in this manner are very unlikely to infect someone else. The third way that infectious microbes can leave our body is through our fecal waste, and the fourth way is through sexual contact.

Simply leaving the body of a person it has infected is only the first step and is not enough for a microbe to spread through a population. It must travel efficiently and spread to other people. Infectious microbes travel and spread in six ways.

Water is the first way that infectious disease can be spread in localized human populations. Approximately one person in five worldwide lacks access to clean and safe water for drinking, cooking, bathing, or watering his or her crops. Contaminated water has resulted in many individual cases and outbreaks of infectious disease. Water is often contaminated with infectious microbes when human sewage or animal waste finds its way into a water supply. Other times, people come into contact with infectious microbes when they enter such aquatic environments as streams or rivers, where certain microbes naturally live.

Notable outbreaks in which the pathogens spread through water include the London cholera outbreak of 1854, and an outbreak of cryptosporidiosis, a gastrointestinal illness, in Milwaukee in 1993. More than 400,000 people became ill in this recent outbreak, making it one of the most severe outbreaks of a water-borne illness ever to occur in the United States.

Contamination of our food supplies is the second outbreak route for infectious diseases. In industrialized nations, most people do not question the safety of their food. Perhaps they fail to realize how far and how long their food may have traveled from farm or field to their dining table. Food can get contaminated at many places along this route. Whether because this route is becoming longer or for other reasons, reports of outbreaks of illnesses caused by microbes entering the body through food are becoming more common.

Some outbreaks result from eating food contaminated by irrigating crops with water containing infectious microbes. Other outbreaks result from changes in the methods used to ship, process, and handle food, which has created new ways for infectious microbes to enter our food supply. For example, food is now often processed in very large batches. This provides increased opportunities for microbes present in small amounts of a particular food to contaminate a very large batch of food. Sometimes the consumers themselves are to blame for food contamination, as when they mishandle or improperly cook the food they eat. One pathogen in the news recently is a strain of *E. coli* bacteria called O157:H7. It caused serious, and occasionally fatal, sickness in people who had eaten infected lettuce or undercooked hamburger, or who had drunk it in unpasteurized apple juice. In all these cases, the *E. coli* contaminating the food came from animal waste. Other recent outbreaks have involved bacteria in the genus *Salmonella* for example, an outbreak involving contaminated ice cream in the Midwestern United States.

Third, infectious disease can reach humans via vectors. Vectors are organisms that carry disease-causing microbes and transmit them from one host to another. The most common vectors are arthropods, particularly insects and ticks.

Because vectors are living things, their populations can evolve adaptations in response to changes in their habitats. Such changes may have natural causes; for example, changes in weather patterns. They may also result from such human actions as deforestation, agriculture, and the introduction of new species. Such evolutionary adaptations by vectors could increase the likelihood of their infecting humans with pathogenic microbes.

Over time too small to involve the evolution of vectors, human actions or natural processes may increase the area in which a particular vector is in contact with people. For example, changes that broaden the habitat of the wild or domestic animal hosts upon which a vector feeds could increase the area of its habitat. It might then occupy territory inhabited by people. Alternatively, humans may expand the territory in which they work, live, or travel, thus encroaching upon territory occupied by one or more species of vector.

Examples of vector-borne outbreaks include the Black Death (or plague) caused by the bacteria *Yersenia pestis*. This microbe infects rodents and is transmitted to humans by the bite of fleas that have previously bitten infected rodents. Another example of a disease transmitted by vectors is malaria. Malaria results from infection by any one of four species of protozoa in the genus *Plasmodium*. These protozoa travel from victim to victim through bites by some species of mosquitoes. Yet another example of a vector-borne disease is Lyme disease. The microbe causing this disease is a bacteria named *Borrelia burgdorferi*, whose usual host is the white-footed mouse (*Peromyscus leucopus*). A species of deer tick (*Ixodes dammini*), whose preferred food source is the white-tailed deer (*Odocoileus virginianus*), sometimes feeds on the mice and then on people, thereby infecting them.

Many microbes travel from person to person via the air, which is the fourth way infectious disease can spread. When we sneeze, cough, or even talk we emit microbe-containing droplets. Some of these droplets are so small that they remain suspended in the air—therefore forming an aerosol—long enough to infect someone else, who perhaps inhales them. This mode of transmission is particularly efficient in such crowded environments as cities, jails, dormitories, and day care centers; here, the droplets do not have far to go to reach another person. Some technological advances—for example, air conditioners, mist machines, humidifiers, and air circulators—have contributed to the distribution of airborne microbes. Perhaps the most famous aerosol-spread outbreaks involve tuberculosis (known historically as consumption, the White Plague, and phthisis); the infecting microbe causing this major human scourge—still the world's leading killer among infectious diseases—is the bacteria *Mycobacterium tuberculosis*, which grows very slowly in the human lung. A recently discovered aerosol-spread disease is the pulmonary disease called Legionnaires' disease. This disease was first recognized in an outbreak in which the pathogen—the bacteria named *Legionella pneumophilia*—spread via a hotel's air-conditioning system.

The fifth way infectious disease can spread is through our blood. Needles are effective at penetrating the protective barrier provided by our skin and allowing physicians to administer necessary drugs and vaccines directly into our bloodstream. In well-developed health-delivery systems, medical personnel use a new needle for every injection, because otherwise small amounts of blood from one patient might contaminate an injection given to another patient. Unfortunately, in many parts of the world resources are so limited that few needles are available. Consequently, these few needles must often be used and reused, often without adequate sterilization between uses. This practice of needle reuse essentially provides a direct path from one person's bloodstream to another's. In many countries, especially in the West, this same direct path from bloodstream to bloodstream often occurs among users of intravenous drugs. In particular, it can occur in those who share unsterilized needles and/or syringes with other users, who may have infectious microbes in their bloodstream. Even intravenous drug users who have injected themselves with drugs only once or twice in their lives have become infected in this manner.

Another source of blood-borne infection has been transfused blood, and administered blood products such as those used to treat hemophilia, that have been inadequately screened for infectious agents. Disease outbreaks involving blood-borne transmission have included hepatitis C, which affects about 3.1 million Americans. Hepatitis C often affects its victims so severely that they require a liver transplant. Before the development of screening tests for the virus causing hepatitis C, most cases of this disease resulted from contaminated blood and blood products, organ transplants, and dialysis and other medical equipment. Today most new hepatitis C infections occur in intravenous drug users, many of whom are young adults.

The last and sixth way the infectious disease can spread is through sexual contact. Unlike most animal species, members of the human species, Homo sapiens, are not beholden to a physiological cycle or reproductive clock that dictates their sexual activity. Humans can freely choose whether to have sex—and when, how often, and with whom. Because of this sexual freedom, microbes that can spread only through sexual contact between two humans have found ways to infect many people. Any time that people engage in "unsafe" sex, microbes that cause a sexually transmitted disease (STD) gain an opportunity to infect a new host. Unsafe sexual practices include having sex with a partner without knowing his or her sexual history, having sex with multiple partners, and having sex with prostitutes or with other people known to have had many sexual partners. Among the important sexually transmitted diseases are syphilis, HIV/AIDS, gonorrhea, and genital herpes.

To explore outbreaks, we pose the following questions:

How have "medical detectives" become critical links in our understanding of diseases and their origins?

PAUL GREENOUGH, Professor of History, University of Iowa, lays out the historical development of epidemiology by looking at how medical detectives work.

What were the keys to unlocking the mystery of a major outbreak of Legionnaires' disease?

In 1976, an outbreak of Legionnaires' disease occurred at a hotel in Philadelphia causing severe illness and multiple deaths. In his essay, JOSEPH E. MCDADE, Associate Director, Laboratory Science, Centers for Disease Control and Prevention, shows how the mystery was solved and discusses the role of environmental factors in disease detection.

There are new tools in the medical detective's arsenal. How do satellite images help scientists predict disease patterns?

JONATHAN PATZ, Director, Program on Health Effects of Global Change, Johns Hopkins University, and GREGORY E. GLASS, Associate Professor of Molecular Microbiology and Immunology at Johns Hopkins University, share their research on the hantavirus outbreak in the Four Corners, to show how satellite imagery can be used to predict disease patterns due to global climate change.

These perspectives on medical detection of infectious disease and the role of government agencies in disease control give insight into how effective response to an outbreak depends as much on individual contribution as it does adequate policy development and enforcement.

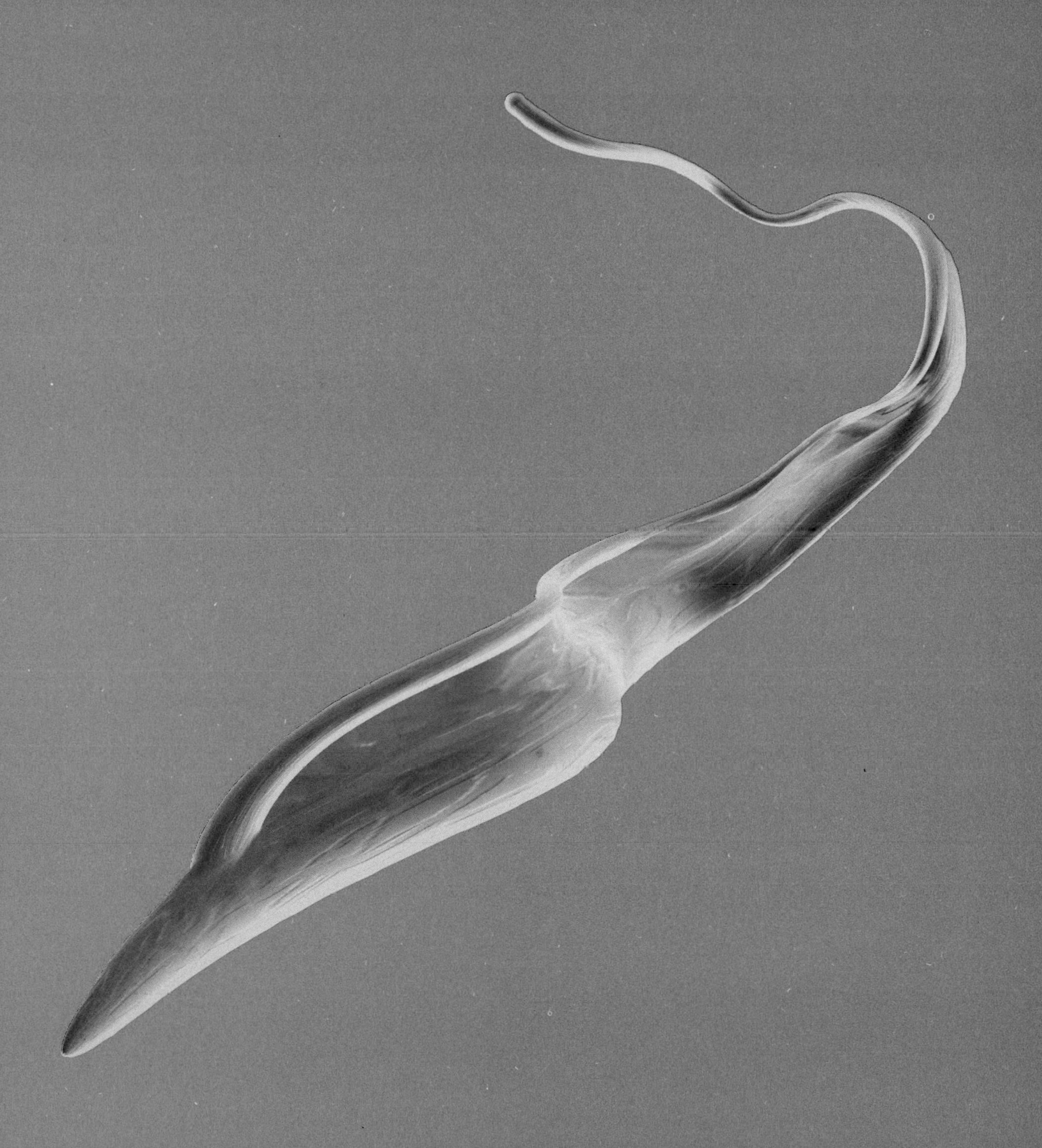

Computer-generated image of a sleeping sickness protozoan (*Trypanosoma brucei*).

Roueché used Sherlock Holmes as a model for how he felt medical epidemiology should be approached. *Shown here:* Basil Rathbone as Sherlock Holmes.

What Is a Medical Detective?

PAUL GREENOUGH

Sherlock Holmes was transformed when he was hot upon such a scent as this...
His nostrils seemed to dilate with a purely animal lust for the chase, and his mind was
so absolutely concentrated upon the matter before him that a question or remark fell
unheeded upon his ears, or, at the most, only provoked a quick, impatient snarl in reply.
Swiftly and silently he made his way along the track which ran through the meadows,
and so by way of the woods to the Boscombe pool. . .

—SIR ARTHUR CONAN DOYLE, *The Boscombe Valley Mystery*

A medical detective is an investigative (or field) epidemiologist who has been trained to collect and interpret disease information under the dynamic conditions of an active outbreak. Medical detectives are usually (but not always) physicians; however, they are doctors without patients. Or rather, whole communities and even whole nations are their patients. There is no Nobel Prize for halting the spread of fatal and crippling infections, but field epidemiologists' service to humanity in recent decades—for example, designing the campaigns that have eradicated smallpox and nearly eradicated polio—suggest to me that such recognition is deserved.

Medical detectives in the United States usually work for the Centers for Disease Control and Prevention (CDC). The CDC first began training field epidemiologists in 1951 when twenty-one newly recruited Public Health Service physicians reported to Atlanta for a two-year course of training at the height of the Korean War. This group was cleverly named the Epidemic Intelligence Service (EIS) to take advantage of Congress's fear that enemy agents might launch biological warfare attacks on the American public. The 1951 EIS class was only the first of nearly fifty subsequent cohorts that have received special training in the detection, analysis, and control of disease. While the EIS has fortunately never had to face an actual biological warfare episode, its officers have made themselves invaluable in natural or spontaneous crises by offering to help state and local health departments in

Paul Greenough is Professor of History at the University of Iowa.

every part of the country. Up to the present time, nearly 2,000 EIS officers have passed through the field epidemiology training program. After serving two years in the Public Health Service, most have fanned out into universities, into state and federal health departments, and into private practice of various kinds. Further, EIS officers have been summoned frequently to assist other countries facing serious outbreaks. Imitation is said to be the sincerest form of flattery, and at least fifteen countries (for example, Mexico, Taiwan, Thailand, and Saudi Arabia) have paid CDC the compliment of setting up independent field epidemiology training programs.

Probably everyone sees the need for an aggressive epidemiological investigation when dangerous diseases strike, but where did the colorful figure of the medical detective come from? To answer this question, we need to know about the first director of the EIS, Alexander Langmuir, a physician and former Johns Hopkins University epidemiologist, who joined CDC in 1949. To recruit EIS officers, Langmuir relied on the "doctor draft" to fill the ranks: in the 1950s, 1960s, and 1970s conscripted physicians could choose between serving in the armed forces or in the Public Health Service, and some of the nation's best-trained and most idealistic young doctors, and a few veterinarians and nurses too, skipped Korea and Vietnam to join the EIS. After 1975, when the draft ended, the reputation of the CDC continued to attract well-qualified applicants into the EIS program. Langmuir recognized that fears of biological warfare would be less compelling after the Korean War ended, and he and others pondered how to convince Congress to continue funding the new organization. It was Langmuir who consciously shaped the public mystique about EIS officers, whom he recast as energetic and insightful medical detectives. Whenever he had the opportunity, Langmuir would draw attention to the rapid response of the EIS leadership to state and local cries for epidemic assistance. Langmuir ordered EIS officers to keep their suitcases packed, and they became famous for arriving at the scene of an outbreak within twenty-four hours of the first phone call. Lightning-like reactions were not exactly the norm within federal bureaucracies, and members of Congress were pleased to tell mayors, governors, and frightened constituents that the EIS was on its way.

It was in late 1952 or early 1953 that Langmuir publicly began to refer to EIS men (and they largely were men for most of Langmuir's tenure) as medical detectives rather than field epidemiologists. Few outside the public health community would have understood in any case what the latter term meant, whereas the notion that a disease investigator was a kind of detective was already familiar in the United States from a best-selling book by Geddes Smith, *Plague on Us*, first published in 1941. In *Plague on Us*, Smith underlined the similarities between an outbreak investigation and a homicide case. His motive for doing so was probably the science writer's perennial problem—how to translate technical matters into familiar idioms. Thus, in a chapter called "Detective Work," which narrated the response of public health investigators to outbreaks of typhoid fever and hemolytic (red-blood-cell–destroying) streptococcus between 1914 and 1936, Smith called the spread of disease the plot; the food and water consumed by stricken patients were suspects; and the laboratory results were clues. Smith's popularizing methods worked, and his book went through three reprintings between 1941 and 1948. Consequently, Langmuir could draw upon an already widespread identification of epidemiology with medical detection in explaining to the public what the EIS was all about.

On January 19, 1953, nearly identical news stories appeared in *Time* and *Newsweek*, both headlined "Disease Detectives" and giving glowing accounts of the first full year of EIS investigations. The *Newsweek* article gave capsule narratives of several EIS investigations, which were drolly labeled "The Case of the Camp Sewage," "The Case of the Carrot Salad," "The Case of the Korean Camper," and "The Case of the Culex Killer." As these titles suggest, EIS investigations at this time were concerned with common source outbreaks; that is, to illnesses within a restricted social group after its members have been exposed to a single noxious substance—archetypally, a spoiled potato salad served at a church picnic. The 1953 stories in *Time* and *Newsweek* set the pattern for further reporting in the press, and EIS officers were regularly referred to as medical or disease detectives—albeit with some effort at variation: they became "medical gumshoes" in *Coronet* (August 1956), "medicine's FBI" in *Reader's Digest* (May 1959), and "sleuths from Atlanta" in *Parent's Magazine* (October 1963). These accounts were short and shallow, but they offered scarce publicity to an agency otherwise rarely in the limelight.

In 1965, a staff writer for the *New Yorker*, Berton Roueché, appeared in Atlanta. Roueché was a skilled writer whose journalism career began under the magazine's legendary founding editor, Harold Ross. At Ross's suggestion, Roueché had been writing up unusual episodes drawn from the files of the New York City Health Department in a *New Yorker* column called "The Annals of Medicine." These were episodes of what he, like Geddes Smith, was happy to call medical detection. In the 1960s, Roueché began to run out of suitable New York City material, and at this time he stumbled upon the EIS. It was he, more than anyone else, who brought depth, moral purpose, and above all intellectual excitement to the image of the medical detective. While the newsweeklies had been content to use the term as a catchy label, Roueché took "medical detection" literally and put the logic-deductive skills of the classical detective at the center of his EIS portraits.

Roueché and Langmuir came to an informal arrangement. Roueché was given access to raw EIS reports, from which he selected the few that met his requirements. In return, CDC and the EIS received publicity in one of the most influential mass journals of American letters. With Langmuir's encouragement, EIS officers spent hours explaining minute details of their investigations to Roueché, reconstructing the logic that they had followed. From this painstaking process came a series of remarkable stories, first appearing in the *New Yorker*, then collected in the late 1960s in a hardcover collection, *The Annals of Epidemiology* (1967). Alex Langmuir wrote a highly favorable introduction to this volume, which, like many of Roueché's books, is still in print.

What are Roueché's stories of medical detection like? They are always based on actual events, and each is focused on a single disease. Although not fictions, they are shaped by the conventions of the classic mystery tale, in which disease replaces crime, and in which the search for the offending pathogen, or for the source of an infection, replaces the hunt for the criminal. The medical investigator, usually a physician or physician-epidemiologist, is cast as an energetic and insightful sleuth who often assists a helpless patient for whom we develop sympathy. The data collected are given to the reader, one by one, as clues in the same sequence in which they are discovered. A sense of crisis hangs over Roueché's narratives, because the patients may worsen or the diseases may spread. Suspense is prolonged in most stories by a digression, wherein Roueché gives the reader enough historical and medical background to appreciate the precise nature of the disease under scrutiny. Sometimes the plots require the investigator to ward off interference from other professionals or to cope with their bungling. Even though the sympathetic patients often die, the stories always end with a complete resolution of the epidemiological issues.

I interviewed Roueché in March 1985 (he died in 1994). He willingly told me that he modeled his stories directly on the Sherlock Holmes tales by Arthur Conan Doyle. Further, he explained that

> My requirements for material are that I want a story of an investigation that is complicated, that has blind alleys and red herrings, and with a denouement that is surprising. So first of all, the investigation has to be successful, these investigators are all heroes, they're winners. An investigation that doesn't get anywhere wouldn't suit my purposes. . . My sense of dramatic structure comes from Sherlock Holmes.

In terms of the Conan Doyle archetypes, the EIS officer regularly plays the role of Holmes. The silly figure of Inspector Lestrade of Scotland Yard, whose narrow outlook and doubting attitude Doyle always satirized, is given to doctors or technicians who miss the diagnostic or epidemiologic clues lying at their feet; while the role of bluff Dr. Watson is taken by Roueché himself. Elsewhere Roueché wrote that

> I attempt to illuminate the method involved in the scientific investigation of epidemic disease. . . My approach is different (from the dry accounts published in public health journals), it is chronological and comprehensive. I reconstruct the investigation as it was actually carried out. I include the floundering up blind allies and stumbles down garden paths. (But) I withhold the source of the trouble as it was withheld from the investigator himself until it is discovered by the evidence. If the form of the narrative seems to resemble that of a classic detective story, it should be remembered that Sir Arthur Conan Doyle derived the Holmesean method from the great Edinburgh diagnostician, Sir Joseph Bell.

As can be seen, Roueché asserted that his stories recreate the experiential world of the epidemiologist. To my knowledge, no one has ever questioned or elaborated Roueché's image of the medical detective on the basis of familiarity with what field epidemiologists actually do. Beyond the fact that CDC has long been content to let Roueché speak for the EIS, there is the curious fact that Roueché's stories have actually been used to recruit and train epidemiologists. The American Public Health Association, for example, noted in 1985 that his stories were "unofficial texts in medical schools and for health professionals in general." In a 1980s survey of former and serving EIS officers, a dozen respondents indicated that Roueché's *New Yorker* stories had attracted them to medicine in general and epidemiology in particular. It is also significant that Roueché was the only non-epidemiologist ever invited to join the prestigious American Epidemiological Society. Roueché not only affected the reading public's perception of field epidemiology, he also deeply impressed field epidemiologists themselves with his captivating, romantic vision of who they are.

While field epidemiologists most often seek to answer questions such as What is the infectious or toxic agent?, How is the agent being transmitted?, How can it be combated?, they rarely do so by themselves, à la Sherlock Holmes. Instead, real field epidemiologists, whom I have been researching for ten years, depend on cooperative relationships with an array of city, county, state, and federal health officials and on laboratory backup provided by bench scientists. Instead of racing along a faint trail with furrowed brow like Sherlock Holmes, they are more likely to pick up the telephone to request hospital administrators for permission to examine patient admission records or to make a public service video on immunization or teen smoking to be broadcast on the six o'clock news. Further, although exotic diseases do sometimes appear, most of the outbreaks investigated by field epidemiologists are well-known entities such as influenza, tuberculosis, AIDS, measles, and hepatitis.

As William Foege, former director of the CDC, says, "When you first hear the sound of hoof beats, think of horses rather than zebras"; that is, consider the obvious and familiar kinds of disease before assuming an exotic, emerging plague. Finally, field epidemiology methods and concerns have gone far beyond the outbreaks of infections that were so important in the 1950s; field epidemiologists now attack problems of environmental contamination, personal violence, product safety and hazards, drug and alcohol abuse, famine, war, natural disasters, and a host of other urgent public health concerns. This is a professional story that needs to be told, but it cannot be told from literary depictions alone.

What is a medical detective?

What are other common images of medical professionals, and how do these images compare to reality?

A Pig from Jersey

BERTON ROUECHÉ

Among those who passed through the general clinic of Lenox Hill Hospital, at Seventy-sixth Street and Park Avenue, on Monday morning, April 6, 1942, was a forty-year-old Yorkville dishwasher whom I will call Herman Sauer. His complaint, like his occupation, was an undistinguished one. He had a stomach ache. The pain had seized him early Sunday evening, he told the examining physician, and although it was not unendurably severe, its persistence worried him. He added that he was diarrheic and somewhat nauseated. Also, his head hurt. The doctor took his temperature and the usual soundings. Neither disclosed any cause for alarm. Then he turned his attention to the manifest symptoms. The course of treatment he chose for their alleviation was unexceptionable. It consisted of a dose of bismuth subcarbonate, a word of dietetic advice, and an invitation to come back the next day if the trouble continued. Sauer went home under the comforting impression that he was suffering from nothing more serious than a touch of dyspepsia.

Sauer was worse in the morning. The pain had spread to his chest, and when he stood up, he felt dazed and dizzy. He did not, however, return to Lenox Hill. Instead, with the inconstancy of the ailing, he made his way to Metropolitan Hospital, on Welfare Island. He arrived there, shortly before noon, in such a state of confusion and collapse that a nurse had to assist him into the examining room. Half an hour later, having submitted to another potion of bismuth and what turned out to be an uninstructive blood count, he was admitted to a general ward for observation. During the afternoon, his temperature, which earlier had been, equivocally, normal, began to rise. When the resident physician reached him on his evening round, it was a trifle over a hundred and three. As is customary in all but the most crystalline cases, the doctor avoided a flat-footed diagnosis. In his record of the case, he suggested three compatible possibilities. One was aortitis, a heart condition caused by an inflammation of the great trunk artery. The others, both of which were inspired by an admission of intemperance that had been wrung from Sauer in the examining room, were cirrhosis of the liver and gastritis due to alcoholism. At the moment, the doctor indicated, the last appeared to be the most likely.

Gastritis, aortitis, and cirrhosis of the liver, like innumerable other ailments, can seldom be repulsed by specific medication, but time is frequently effective. Sauer responded to neither. His fever held and his symptoms multiplied. He itched all over, an edema sealed his eyes, his voice faded and failed, and the seething pains in his chest and abdomen advanced to his arms and legs. Toward the end of the week, he sank into a stony, comalike apathy. Confronted by this disturbing decline, the house physician reopened his mind and reconsidered the evidence. His adaptability was soon rewarded. He concluded that he was up against an acute and, to judge from his patient's progressive dilapidation, a peculiarly rapacious infection. It was an insinuating notion, but it had one awkward flaw. The white-blood-cell count is a reliable barometer of infection, and Sauer's count had been entirely normal. On Wednesday, April 15, the doctor requested that another count be made. He did not question the accuracy of the original test, but the thought had occurred to him that it might have been made prematurely. The report from the laboratory was on his desk when he reached the hospital the following day. It more than confirmed his hunch. It also relieved him simultaneously of both uncertainty and hope. Sauer's white count was morbidly elevated by a preponderance of eosinophiles, a variety of cell that is produced by several potentially epidemic diseases but just one as formidably dishevelling as the case in question. The doctor put down the report and called the hospital superintendent's office. He asked the clerk who answered the phone to inform the Department of Health, to which the appearance of any disease of an epidemiological nature must be promptly communicated, that he had just uncovered a case of trichinosis.

The cause of trichinosis is a voracious endoparasitic worm, *Trichinella spiralis*, commonly called trichina, that lodges in the muscle fibres of an animal host. It enters the host by way of the alimentary canal, and in the intestine produces larvae that penetrate the intestinal walls to enter the blood stream. The worm is staggeringly prolific, and it has been known to remain alive, though quiescent, in the body of a surviving victim for thirty-one years. In general, the number of trichinae that succeed in reaching the muscle determines the severity of an attack. As such parasitic organisms go, adult trichinae are relatively large, the males averaging one-twentieth of an inch in length and the females about twice that. The larvae are less statuesque. Pathologists have found as many as twelve hundred of them encysted in a single gram of tissue. Numerous animals, ranging in size from the mole to the hippopotamus, are hospitable to the trichina, but it has a strong predilection for swine and man. Man's only important source of infection is pork. The disease is perpetuated in swine by the practice common among hog raisers of using garbage, some of which inevitably contains trichinous meat, for feed. Swine have a high degree of tolerance for the trichina, but man's resistive powers are feeble. In 1931, in Detroit, a man suffered a violent seizure of trichinosis as a result of merely eating a piece of bread buttered with a knife that had been used to slice an infested sausage. The hog from which the sausage was made had appeared to be in excellent health. Few acute afflictions are more painful than trichinosis, or more prolonged and debilitating. Its victims are occasionally prostrated for many months, and relapses after apparent recoveries are not uncommon. Its mortality rate is disconcertingly variable. It is usually around six percent, but in some outbreaks nearly a third of those stricken have died, and the recovery of a patient from a full-scale attack is almost unheard of. Nobody is, or can be rendered, immune to trichinosis. Also, there is no specific cure. In the opinion of most investigators, it is far from likely that one will ever be found.

They are persuaded that any therapeutic agent potent enough to kill a multitude of embedded trichinae would probably kill the patient, too.

Although medical science is unable to terminate, or even lessen the severity of, an assault of trichinosis, no disease is easier to dodge. There are several dependable means of evasion. Abstention from pork is, of course, one. It is also the most venerable, having been known, vigorously recommended, and widely practiced for at least three thousand years. Some authorities, in fact, regard the Mosaic proscription of pork as the pioneering step in the development of preventive medicine. However, since the middle of the nineteenth century, when the cause and nature of trichinosis were illuminated by Sir James Paget, Rudolf Virchow, Friedrich Albert von Zenker, and others, less ascetic safeguards have become available. The trichinae are rugged but not indestructible. It has been amply demonstrated that thorough cooking (until the meat is bone-white) will make even the wormiest pork harmless. So will refrigeration at a maximum temperature of five degrees for a minimum of twenty days. So, just as effectively, will certain scrupulous methods of salting, smoking, and pickling.

Despite this abundance of easily applied defensive techniques, the incidence of trichinosis has not greatly diminished over the globe in the past fifty or sixty years. In some countries, it has even increased. The United States is one of them. Many epidemiologists are convinced that this country now leads the world in trichinosis. It is, at any rate, a major health problem here. According to a compendium of recent autopsy studies, approximately one American in five has at some time or another had trichinosis, and it is probable that well over a million are afflicted with it every year. As a considerable source of misery, it ranks with tuberculosis, syphilis, and undulant fever. It will probably continue to be one for some time to come. Its spread is almost unimpeded. A few states, New York among them, have statutes prohibiting the feeding of uncooked garbage to swine, but nowhere is a very determined effort made at enforcement, and the Bureau of Animal Industry of the United States Department of Agriculture, although it assumes all pork to be trichinous until proved otherwise, requires packing houses to administer a prophylactic freeze to only those varieties of the meat—frankfurters, salami, prosciutto, and the like—that are often eaten raw. Moreover, not all processed pork comes under the jurisdiction of the Department. At least a third of it is processed under local ordinances in small, neighborhood abattoirs beyond the reach of the Bureau, or on farms. Nearly two per cent of the hogs slaughtered in the United States are trichinous.

Except for a brief period around the beginning of this century, when several European countries refused, because of its dubious nature, to import American pork, the adoption of a less porous system of control has never been seriously contemplated here. One reason is that it would run into money. Another is that, except by a few informed authorities, it has always been considered unnecessary. Trichinosis is generally believed to be a rarity. This view, though hallucinated, is not altogether without explanation. Outbreaks of trichinosis are seldom widely publicized. They are seldom even recognized. Trichinosis is the chameleon of diseases. Nearly all diseases are anonymous at onset, and many tend to resist identification until their grip is well established, but most can eventually be identified by patient scrutiny. Trichinosis is occasionally impervious to bedside detection at any stage. Even blood counts sometimes inexplicably fail to reveal its presence at any stage in its development. As a diagnostic deadfall, it is practically unique. The number and variety of ailments with which it is more or less commonly confused approach the encyclopedic. They include arthritis, acute alcoholism, conjunctivitis, food poisoning, lead poisoning, heart disease, laryngitis, mumps, asthma, rheumatism, rheumatic fever, rheumatic myocarditis, gout, tuberculosis, angioneurotic edema, dermatomyositis, frontal sinusitis, influenza, nephritis, peptic ulcer, appendicitis, cholecystitis, malaria, scarlet fever, typhoid fever, paratyphoid fever, undulant fever, encephalitis, gastroenteritis, intercostal neuritis, tetanus, pleurisy, colitis, meningitis, syphilis, typhus, and cholera. It has even been mistaken for beriberi. With all the rich inducements to error, a sound diagnosis of trichinosis is rarely made, and the diagnostician cannot always take much credit for it. Often, as at Metropolitan Hospital that April day in 1942, it is forced upon him.

The report of the arresting discovery at Metropolitan reached the Health Department on the morning of Friday, April 17. Its form was conventional—a postcard bearing a scribbled name, address, and diagnosis—and it was handled with conventional dispatch. Within an hour, Dr. Morris Greenberg, who was then chief epidemiologist of the Bureau of Preventable Diseases and is now its director, had put one of his fleetest agents on the case, a field epidemiologist named Lawrence Levy. Ten minutes after receiving the assignment, Dr. Levy was on his way to the hospital, intent on tracking down the source of the infection, with the idea of alerting the physicians of other persons who might have contracted the disease along with Sauer. At eleven o'clock, Dr. Levy walked into the office of the medical superintendent at Metropolitan. His immediate objective was to satisfy himself that Sauer was indeed suffering from trichinosis. He was quickly convinced. The evidence of the eosinophile count was now supported in the record by more graphic proof. Sauer, the night before, had undergone a biopsy. A sliver of muscle had been taken from one of his legs and examined under a microscope. It teemed with *Trichinella spiralis*. On the basis of the sample, the record noted, the pathologist who made the test estimated the total infestation of trichinae at upward of twelve million. A count of over five million is almost invariably lethal. Dr. Levy returned the dossier to the file. Then, moving on to his more general objective, he had a word with the patient. He found him bemused but conscious. Sauer appeared at times to distantly comprehend what was said to him, but his replies were faint and rambling and mostly incoherent. At the end of five minutes, Dr. Levy gave up. He hadn't learned much, but he had learned something, and he didn't have the heart to go on with his questioning. It was just possible, he let himself hope, that he had the lead he needed. Sauer had mentioned the New York Labor Temple, a German-American meeting-and-banquet hall on East Eighty-fourth Street, and he had twice uttered the word "*Schlachtfest*." A *Schlachtfest*, in Yorkville, the Doctor knew, is a pork feast.

Before leaving the hospital, Dr. Levy telephoned Dr. Greenberg and dutifully related what he had found out. It didn't take him long. Then he had a sandwich and a cup of coffee and headed for the Labor Temple, getting there at a little past one. It was, and is, a shabby yellow-brick building of six stories, a few doors west of Second Avenue, with a high, ornately balustraded stoop and a double basement. Engraved on the façade, just above the entrance, is a maxim: "Knowledge Is Power." In 1942, the Temple was owned and operated, on a non-profit basis, by the Workmen's Educational

Association; it has since been acquired by private interests and is now given over to business and light manufacturing. A porter directed Dr. Levy to the manager's office, a cubicle at the end of a dim corridor flanked by meeting rooms. The manager was in, and, after a spasm of bewilderment, keenly cooperative. He brought out his records and gave Dr. Levy all the information he had. Sauer was known at the Temple. He had been employed there off and on for a year or more as a dishwasher and general kitchen helper, the manager related. He was one of a large group of lightly skilled wanderers from which the cook was accustomed to recruit a staff whenever the need arose. Sauer had last worked at the Temple on the nights of March 27 and March 28. On the latter, as it happened, the occasion was a *Schlachtfest*.

Dr. Levy, aware that the incubation period of trichinosis is usually from seven to fourteen days and that Sauer had presented himself at Lenox Hill on April 6, motioned to the manager to continue. The *Schlachtfest* had been given by the Hindenburg Pleasure Society, an informal organization whose members and their wives gathered periodically at the Temple for an evening of singing and dancing and overeating. The arrangements for the party had been made by the secretary of the society —Felix Lindenhauser, a name which, like those of Sauer and the others I shall mention in connection with the *Schlachtfest*, is a fictitious one. Lindenhauser lived in St. George, on Staten Island. The manager's records did not indicate where the pork had been obtained. Probably, he said, it had been supplied by the society. That was frequently the case. The cook would know, but it was not yet time for him to come on duty. The implication of this statement was not lost on Dr. Levy. Then the cook, he asked, was well? The manager said that he appeared to be. Having absorbed this awkward piece of information, Dr. Levy inquired about the health of the others who had been employed in the kitchen on the night of March 28. The manager didn't know. His records showed, however, that, like Sauer, none of them had worked at the Temple since that night. He pointed out that it was quite possible, of course, that they hadn't been asked to. Dr. Levy noted down their names—Rudolf Nath, Henry Kuhn, Frederick Kreisler, and William Ritter—and their addresses. Nath lived in Queens, Kreisler in Brooklyn, and Kuhn and Ritter in the Bronx. Then Dr. Levy settled back to await the arrival of the cook. The cook turned up at three, and he, too, was very cooperative. He was feeling fine, he said. He remembered the *Schlachtfest*. The pig, he recalled, had been provided by the society. Some of it had been ground up into sausage and baked. The rest had been roasted. All of it had been thoroughly cooked. He was certain of that. The sausage, for example, had been boiled for two hours before it was baked. He had eaten his share of both. He supposed that the rest of the help had, too, but there was no knowing. He had neither seen nor talked to any of them since the night of the feast. There had been no occasion to, he said.

Dr. Levy returned to his office, and sat there for a while in meditation. Presently, he put in a call to Felix Lindenhauser, the secretary of the society, at his home on Staten Island. Lindenhauser answered the telephone. Dr. Levy introduced himself and stated his problem. Lindenhauser was plainly flabbergasted. He said he was in excellent health, and had been for months. His wife, who had accompanied him to the *Schlachtfest*, was also in good health. He had heard of no illness in the society. He couldn't believe that there had been anything wrong with that pork. It had been delicious. The pig had been obtained by two members of the society, George Muller and Hans Breit, both of whom lived in the Bronx. They had bought it from a farmer of their acquaintance in New Jersey. Lindenhauser went on to say that there had been twenty-seven people at the feast, including himself and his wife. The names and addresses of the company were in his minute book. He fetched it to the phone and patiently read them off as Dr. Levy wrote them down. If he could be of any further help, he added as he prepared to hang up, just let him know, but he was convinced that Dr. Levy was wasting his time. At the moment, Dr. Levy was almost inclined to agree with him.

Dr. Levy spent an increasingly uneasy weekend. He was of two antagonistic minds. He refused to believe that Sauer's illness was not in some way related to the *Schlachtfest* of the Hindenburg Pleasure Society. On the other hand, it didn't seem possible that it was. Late Saturday afternoon, at his home, he received a call that increased his discouragement, if not his perplexity. It was from his office. Metropolitan Hospital had called to report that Herman Sauer was dead. Dr. Levy put down the receiver with the leaden realization that, good or bad, the *Schlachtfest* was now the only lead he would ever have.

On Monday, Dr. Levy buckled heavily down to the essential but unexhilarating task of determining the health of the twenty-seven men and women who had attended the *Schlachtfest*. Although his attitude was half-hearted, his procedure was methodical, unhurried, and objective. He called on and closely examined each of the guests, including the Lindenhausers, and from each procured a sample of blood for analysis in the Health Department laboratories. The job, necessarily involving a good deal of leg work and many evening visits, took him the better part of two weeks. He ended up, on April 30th, about equally reassured and stumped. His findings were provocative but contradictory. Of the twenty-seven who had feasted together on the night of March 28, twenty-five were in what undeniably was their normal state of health. Two, just as surely, were not. The exceptions were George Muller and Hans Breit, the men who had provided the pig. Muller was at home and in bed, suffering sorely from what his family physician had uncertainly diagnosed as some sort of intestinal upheaval. Breit was in as bad a way, or worse, in Fordham Hospital. He had been admitted there for observation on April 10. Several diagnoses had been suggested, including rheumatic myocarditis, pleurisy, and grippe, but none had been formally retained. The nature of the two men's trouble was no mystery to Dr. Levy. Both, as he was subsequently able to demonstrate, had trichinosis.

On Friday morning, May 1, Dr. Levy returned to the Bronx for a more searching word with Muller. Owing to Muller's debilitated condition on the occasion of Dr. Levy's first visit, their talk had been brief and clinical in character. Muller, who was now up and shakily about, received him warmly. Since their meeting several days before, he said, he had been enlivening the tedious hours of illness with reflection. A question had occurred to him. Would it be possible, he inquired, to contract trichinosis from just a few nibbles of raw pork? It would, Dr. Levy told him. He also urged him to be more explicit. Thus encouraged, Muller displayed an unexpected gift for what appeared to be total recall. He leisurely recounted to Dr. Levy that he and Breit had bought the pig from a farmer who owned a place near Midvale, New Jersey. The farmer had killed and dressed the animal, and they had delivered the carcass to the Labor Temple kitchen on the evening of March 27. That, however, had been only part of their job. Not wishing to trouble the cook and his helpers, who were otherwise occupied, Muller and Breit had then set about preparing the sausage for the feast. They were both experienced amateur sausage makers, he said, and explained the process—grinding, maccrating, and seasoning—in laborious detail. Dr. Levy began to fidget. Naturally, Muller presently went on, they had been obliged to sample their work. There was no other way to make sure that the meat was properly seasoned. He had taken perhaps two or three little nibbles. Breit, who had a heartier taste for raw pork, had probably eaten a trifle more. It was hard to believe, Muller said, that so little—just a pinch or two—could cause such misery. He had thought his head would split, and the pain in his legs had been almost beyond endurance. Dr. Levy returned him sympathetically to the night of March 27. They had finished with the sausage around midnight, Muller remembered. The cook had departed by then, but his helpers were still at work. There had been five of them. He didn't know their names, but he had seen all or most of them again the next night, during the feast. Neither he nor Breit had given them any of the sausage before they left. But it was possible, of course, since the refrigerator in which he and Breit had stored the meat was not, like some, equipped with a lock...Dr. Levy thanked him, and moved rapidly to the door.

Dr. Levy spent the rest of the morning in the Bronx. After lunch, he hopped over to Queens. From there, he made his way to Brooklyn. It was past four by the time he got back to his office. He was hot and gritty from a dozen subway journeys, and his legs ached from pounding pavements and stairs and hospital corridors, but he had tracked down and had a revealing chat with each of Sauer's kitchen colleagues, and his heart was light. Three of them—William Ritter, Rudolf Nath, and Frederick Kreisler—were in hospitals. Ritter was at Fordham, Nath at Queens General, and Kreisler at the Coney Island Hospital, not far from his home in Brooklyn. The fourth member of the group, Henry Kuhn, was sick in bed at home. All were veterans of numerous reasonable but incorrect diagnoses, all were in more discomfort than danger, and all, it was obvious to Dr. Levy's unclouded eye, were suffering from trichinosis. Its source was equally obvious. They had prowled the icebox after the departure of Muller and Breit, come upon the sausage meat, and cheerfully helped themselves. They thought it was hamburger.

Before settling down at his desk to compose the final installment of his report, Dr. Levy looked in on Dr. Greenberg. He wanted, among other things, to relieve him of the agony of suspense. Dr. Greenberg gave him a chair, a cigarette, and an attentive ear. At the end of the travelogue, he groaned. "Didn't they even bother to cook it?" he asked.

"Yes, most of them did," Dr. Levy said. "They made it up into patties and fried them. Kuhn cooked his fairly well. A few minutes, at least. The others liked theirs rare. All except Sauer. He ate his raw."

"Oh," Dr. Greenberg said.

"Also," Dr. Levy added, "he ate two."

"A Pig from New Jersey," from *The Medical Detectives*, by Berton Roueché, Truman Talley Books, New York.

Profile

Bonnie Smoak: A Soldier in the War Against Infectious Disease

If someone asked you how to go about launching a career in medicine, joining the U.S. Army probably would not be the first thing you'd think of. But that's exactly what Lieutenant Colonel Bonnie Smoak, M.D., Ph.D., M.P.H., did.

She had already earned a doctorate in exercise physiology when she decided to go to medical school. By then, she had been a higher-education student for the better part of fifteen years, so funds to pay for medical school were in short supply. Her solution was to enlist in the army. Uncle Sam paid her medical school tuition at the University of Chicago, and Bonnie Smoak has been returning the favor ever since.

After earning her M.D. in 1985, she joined the faculty and research staff at the Uniformed Services University of Health Sciences, which trains physicians for the armed services. Her main focus was exercise physiology, but the more she learned about the role the military had played in infectious disease research, the more intrigued with the subject she became.

She followed her curiosity to the Harvard School of Public Health, where she earned an M.P.H. degree, and then to a residency in Preventive Medicine at the Walter Reed Army Institute of Research (WRAIR). WRAIR is part of the Army's Medical Research and Materiel Command. It was named in honor of the American bacteriologist and army surgeon who uncovered the cause of yellow fever, a mosquito-borne virus, and made significant discoveries in the field of typhoid fever. Although Walter Reed's work was conducted in his capacity as a military physician and for the benefit of American troops, his contributions have added to medical knowledge and improved the health of generations of civilians.

Soldier-scientist Bonnie Smoak follows that tradition, working with the military to make a difference to all humankind. In the decade during which she has served her country as an infectious disease epidemiologist, she has studied drug-resistant malaria in Somalia, dengue fever in Haiti, tick typhus in Botswana, and hemorrhagic fevers in Kenya. In each case, her assignment grew out of an American military presence in the region, but has had an impact on the health of civilian populations in that country and elsewhere in the world.

One of the highest priorities of the military is the health of its service members. Before troops are deployed anywhere in the world, infectious disease threats are identified and preventive measures taken. During the military operation, health workers monitor the troops and outbreaks of infectious disease are investigated to prevent further casualties. Even after troops return home, their health status continues to be monitored. For example, in 1994, during Operation Uphold Democracy in Haiti, some American soldiers were being infected by the mosquito-borne dengue virus. Public health authorities at home were notified about the potential for additional cases of this tropical disease in returning soldiers.

"Military forces deployed around the world are exposed to different diseases and environmental threats than people living in America," Dr. Smoak explains. "Military researchers address medical problems of the armed services, seeking to discover or identify infectious disease agents, then to develop protective measures, such as a vaccine or drug. Despite focusing on military problems, the results of our research are applicable to civilians." Examples include the development of vaccines for hepatitis A, a viral infection that damages the liver, and Japanese encephalitis, a mosquito-borne viral infection of the brain. In addition, researchers at the U.S. Army Medical Research Institute for Infectious Diseases (USAMRIID) used the vast virus library maintained by Department of Defense to help solve the mystery of the Sin Nombre virus outbreak in Four Corners in 1993.

Aside from research, she points out, "The U.S. military frequently participates in humanitarian efforts, responding to medical emergencies in other countries," often in collaboration with the U.S. State Department, local ministries of health, WHO, and other international health agencies, and non-governmental organizations.

Dr. Smoak is now Chief of Epidemiology and Disease Surveillance at the U.S. Army Medical Research Laboratory in Kenya, one of four special foreign activities of WRAIR. The others are in Germany, Brazil, and Thailand. Her main duty is as Director of the Department of Defense Global Emerging Infections System in East Africa, a keystone of the international effort to detect and monitor emerging pathogens and improve response capabilities when outbreaks occur.

"In collaboration with my international colleagues, I am establishing surveillance systems to monitor drug resistance in malaria parasites and enteric [intestinal] pathogens in East Africa," she says. "We are also investigating the infectious causes of fevers among indigenous populations in hopes of identifying new organisms that cause disease." In the past few years alone, she has investigated outbreaks of cholera, Rift Valley fever, and life-threatening dysentery caused by the enterobacterium *Shigella dysenteriae*.

It's a long way from Washington, D.C., where she was born, to Nairobi, Kenya, and from exercise physiology to Rift Valley fever, but for Bonnie Smoak, what started out as a gambit for medical school tuition has turned into a lifetime of service, in the American military and in the battle against infectious disease.

Lieutenant Colonel Bonnie Smoak in the clinic.

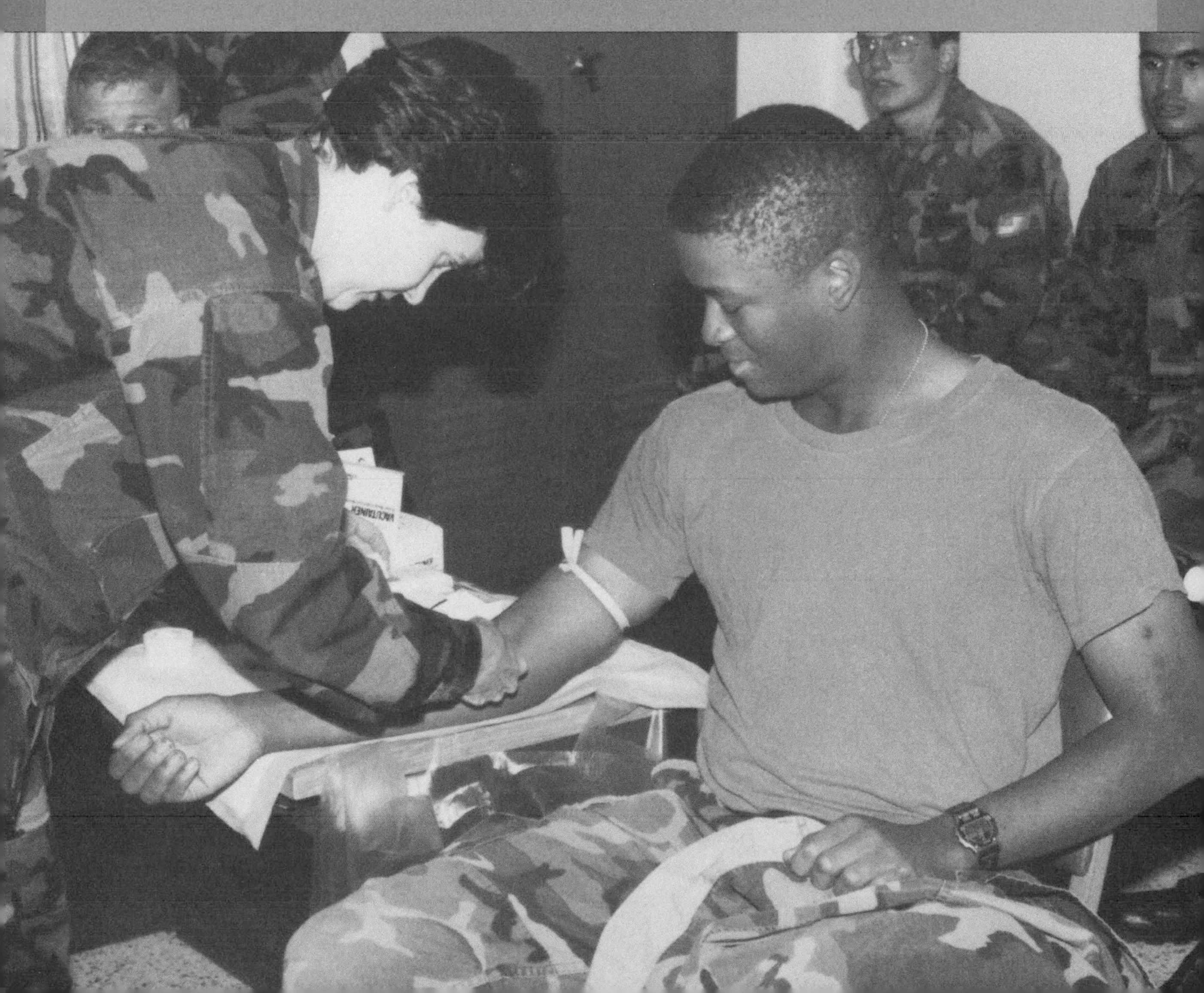

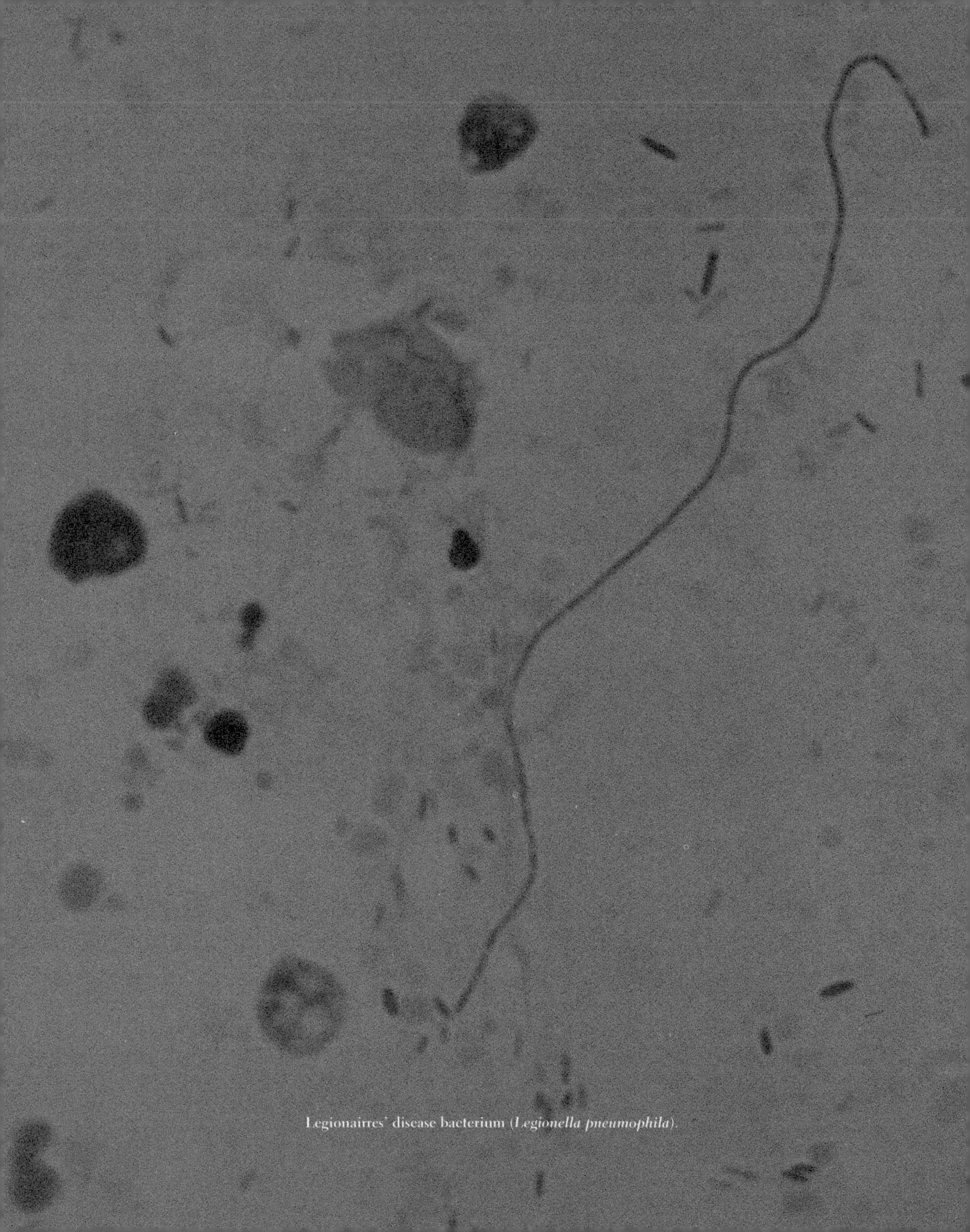

Legionairres' disease bacterium (*Legionella pneumophila*).

The Ecology of Outbreaks: Discovery of *Legionella pneumophila*

JOSEPH E. MCDADE

The Centers for Disease Control and Prevention (CDC) in Atlanta, a division of the U.S. Department of Health and Human Services, has a well-earned reputation for protecting the health of the public. For decades, CDC has been the primary agency that investigates infectious disease outbreaks in the United States and around the world. Usually, it identifies the cause of these outbreaks in a few days or, at most, a few weeks. Sometimes, determining the cause of outbreaks can be much more difficult. When I joined CDC in 1975 as a laboratory scientist, I never thought I would be involved in an outbreak investigation; however, I soon found myself amid one of the most publicized epidemics in CDC's history.

In July 1976, during America's Bicentennial celebration, the Pennsylvania branch of the American Legion held its annual convention at the Bellevue-Stratford Hotel in Philadelphia. The convention appeared to progress normally. However, after the attendees returned home, many experienced a puzzling illness, characterized by fever, coughing, and pneumonia. By early August, the Pennsylvania Department of Health had received reports of this illness from around the state; clearly, a serious disease outbreak had occurred among persons who had attended the convention. The press called the illness Legionnaires' disease because the first reported cases occurred among those military veterans.

The CDC sent twenty-three Epidemic Intelligence Service officers to Pennsylvania to work with dozens of local public health workers in Pennsylvania to learn the cause of the outbreak. Within a short time, investigators learned that 221 people had become ill and that thirty-four had died of pneumonia or its complications. Not all of the sick people were veterans, but everyone who became ill had either stayed in or been near the Bellevue-Stratford hotel. Contacts of the patients remained well, showing that the disease did not spread from

Joseph E. McDade is the Deputy Director of the National Center for Infectious Diseases, Centers for Disease Control and Prevention, Atlanta, Georgia.

person to person. Investigators quickly ruled out the possibility that contaminated food or drinking water caused the disease. One clue suggested that the disease might have spread through the air: ill persons had spent more time in the lobby of the hotel or on the sidewalk in front of the hotel than persons who stayed well.

The symptoms of the disease were not unique to a particular illness and could have been caused by a variety of toxic agents or microorganisms. Tissue specimens, collected at autopsies from deceased patients, and blood specimens from living patients were sent to CDC for laboratory tests that might identify the cause. However, extensive testing, conducted over several months, failed to identify the consistent presence of any toxic chemicals or microorganisms that might have caused the illness. In December, the cause of the outbreak was still unknown.

My task was to learn whether the cause of Q fever was also the cause of Legionnaires' disease. Q fever is a mild type of pneumonia caused by a certain type of bacteria, called rickettsiae, which are found in the milk or placentas of infected domestic animals, especially cows. Q fever rickettsiae become airborne when infected animals give birth. Persons who work with domestic livestock, such as farmers and employees of meat-packing plants, are at greatest risk for inhaling these rickettsiae and contracting Q fever. Because downtown Philadelphia housed no domestic livestock, it seemed highly unlikely that the outbreak was Q fever. Nonetheless, I did a series of tests to rule out Q fever, just to be certain.

Q fever rickettsiae can only grow in living cells—for example, in experimental animals or tissue cultures. I attempted to isolate Q fever rickettsiae by inoculating lung tissue from deceased Legionnaires' patients into guinea pigs. The animals became ill several days later, but I did not observe any rickettsiae in stained smears of guinea pig tissues. I did notice in these tissues a rare, rod-shaped bacterium, which was larger than Q fever rickettsiae; however, I could not recover it when I attempted to grow it on synthetic media in glass culture vessels. This was extremely puzzling; I could see a bacterium, but I was unable to cultivate it. At first, I did not follow up on this observation, because I thought the bacterium was an insignificant contaminant introduced during tissue processing. However, by December, when the cause of the outbreak still had not been found, I began to wonder whether the bacterium had more significance.

I tried another approach to growing this fastidious bacterium. I retrieved the suspensions of guinea pig tissue that I had saved from my initial experiment and inoculated them into embryo-containing hens' eggs: these provide a favorable environment for growing fastidious bacteria. The embryos died several days later, and the rod-shaped bacterium was clearly visible in smears of the yolk sac membranes. I repeated the isolation procedure with other pieces of autopsy tissue from patients and recovered the same bacterium. The question that remained was whether the bacterium was the cause of the outbreak or just an experimental artifact.

My colleagues and I tested the serum specimens of Legionnaires' disease patients who survived the infection to see if they contained antibodies to the newly isolated bacterium. If a patient's antibody level rose during their convalescence, then it was very likely that this microorganism had caused the illness. Test results showed that more than ninety percent of the patients had high levels of antibodies to the bacterium: we had found the cause of the Legionnaires' disease outbreak.

Several scientists at CDC who worked on the Legionnaires' disease outbreak had also worked on two earlier outbreaks of pneumonia and respiratory illness many years earlier: at St. Elizabeth's Hospital in Washington, D.C., in 1965, and in Pontiac, Michigan, in 1968. The cause of those outbreaks was still unknown. However, serum specimens from patients in those outbreaks remained in freezers at CDC. When we tested those serum specimens, we found that most of them had high levels of antibodies to the bacterium, showing that it had also caused these earlier pneumonia outbreaks. Eventually, scientists at CDC could grow the bacterium on synthetic growth medium, which made it easier to study. Extensive characterization of the new bacterium showed that it was indeed a new species. It was named *Legionella pneumophila*. Other species of *Legionella* have subsequently been isolated, but only some of them cause disease in humans. Additionally, we now know that *L. pneumophila* has several different subtypes and that some subtypes are associated with human illness more frequently than others.

Determining how *Legionella* survives and spreads to humans proved extremely interesting. Most infectious agents that cause epidemics of

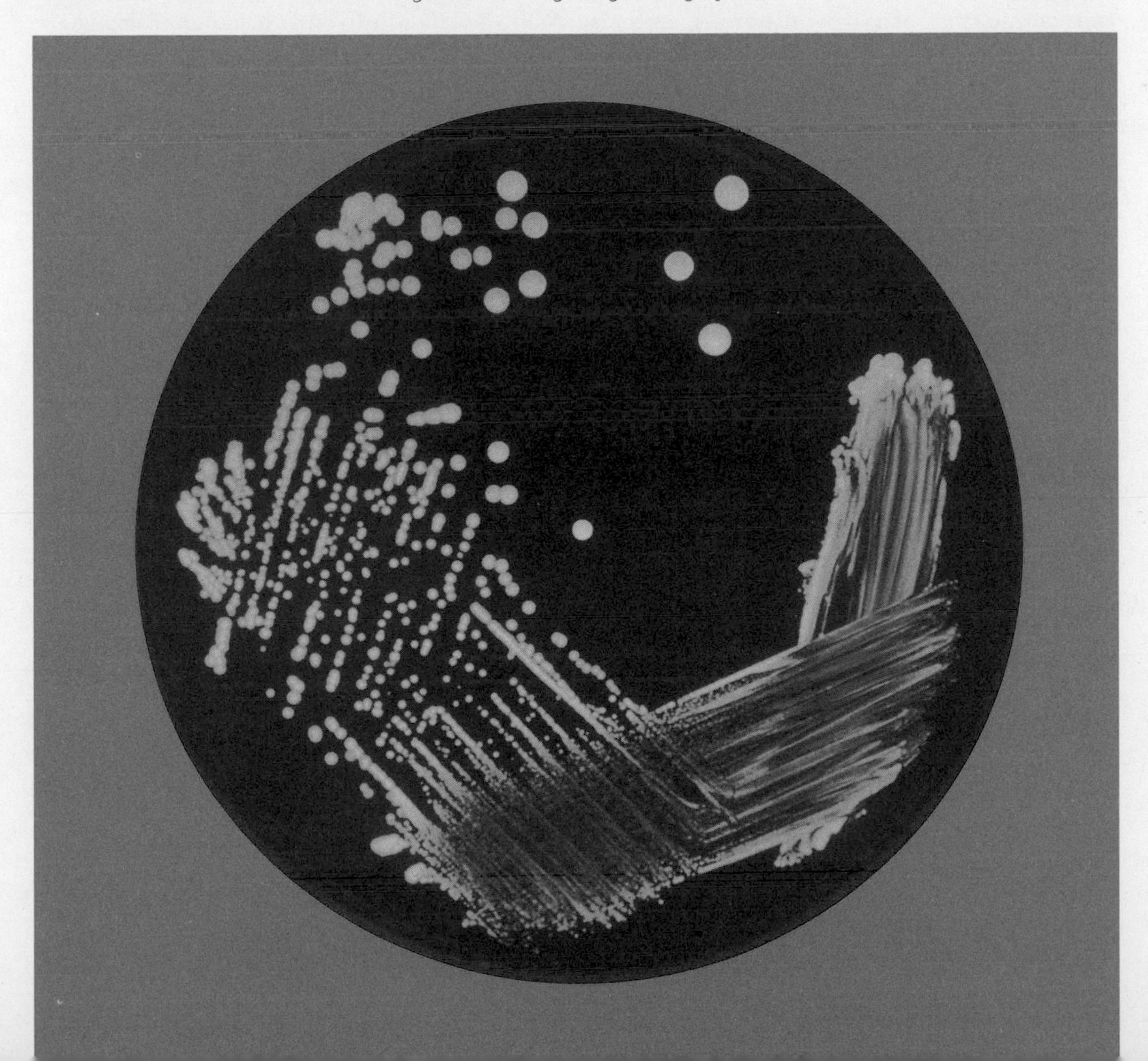
Legionella bacteria growing on an agar plate.

pneumonia, but do not spread from person to person, spread through the air from a characteristic ecological niche. A common example is the fungus *Histoplasma capsulatum*, which lives in the soil and causes disease when contaminated dust is stirred up and inhaled. It seemed likely that *L. pneumophila* might be found in such an inanimate environment. This hypothesis gained support from laboratory studies showing that the bacterium could survive for more than a year in tap water.

Scientists could test this hypothesis when additional outbreaks of Legionnaires' disease occurred after the Philadelphia episode. In several outbreaks, the water in the cooling towers or evaporative condensers of air-conditioning systems appeared to be the source of infection. The design of the air-conditioning systems provided a mechanism for human exposure to *L. pneumophila*. For example, in an evaporative condenser, water is sprayed over metal coils containing the refrigerant; the refrigerant, which has been heated through its rapid compression and condensation at this stage of the refrigeration cycle, is cooled directly by the evaporation of the water. Minute droplets of water form a train of drift, which carries whatever is in the water, including any bacteria that may be present, a considerable distance. In several outbreaks, patients had been exposed to drift, and in each instance *L. pneumophila* was isolated from water in the cooling device.

Outbreaks of Legionnaires' disease subsequently have been associated with contaminated water in whirlpool baths, showers, respiratory care equipment, decorative fountains, and even produce-misting devices in supermarkets. However, investigators have recovered *Legionella* species from these sources even when no outbreaks have occurred. An outbreak of pneumonia requires a combination of factors: virulent strains of *Legionella* must be present in water, aerosolized droplets of contaminated water must enter the air, and susceptible persons must inhale the aerosol. Recent studies have also shown that *Legionella* species are common in streams and other natural sources of fresh water, where they have undoubtedly existed for thousands of years. Some species of *Legionella* grow extremely well in fresh water protozoa. Scientists speculate that adaptation to growth in protozoa may have given *Legionella* organisms the capability to proliferate in certain types of cells in human lungs.

In sum, changing technology and the existence of new ecological niches are responsible for outbreaks of Legionnaires' disease. The development of large air-conditioning systems and other devices that aerosolize water, in close association with large human populations, has provided an ideal environment for *Legionella* infections. Otherwise, *Legionella* would reside harmlessly in protozoa in local streams.

How was Legionnaires' disease discovered in the Philadelphia outbreak?

What was so unusual about the transmission?

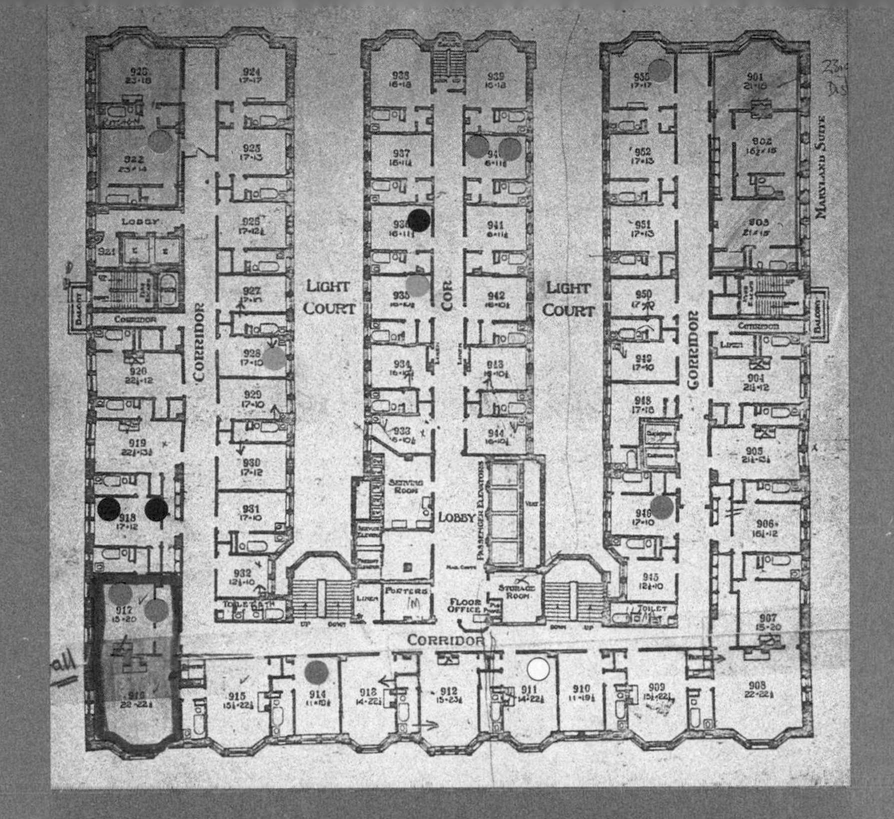

Floor plans of the Bellevue-Stratford Hotel where Legionnaires' disease broke out in 1976.

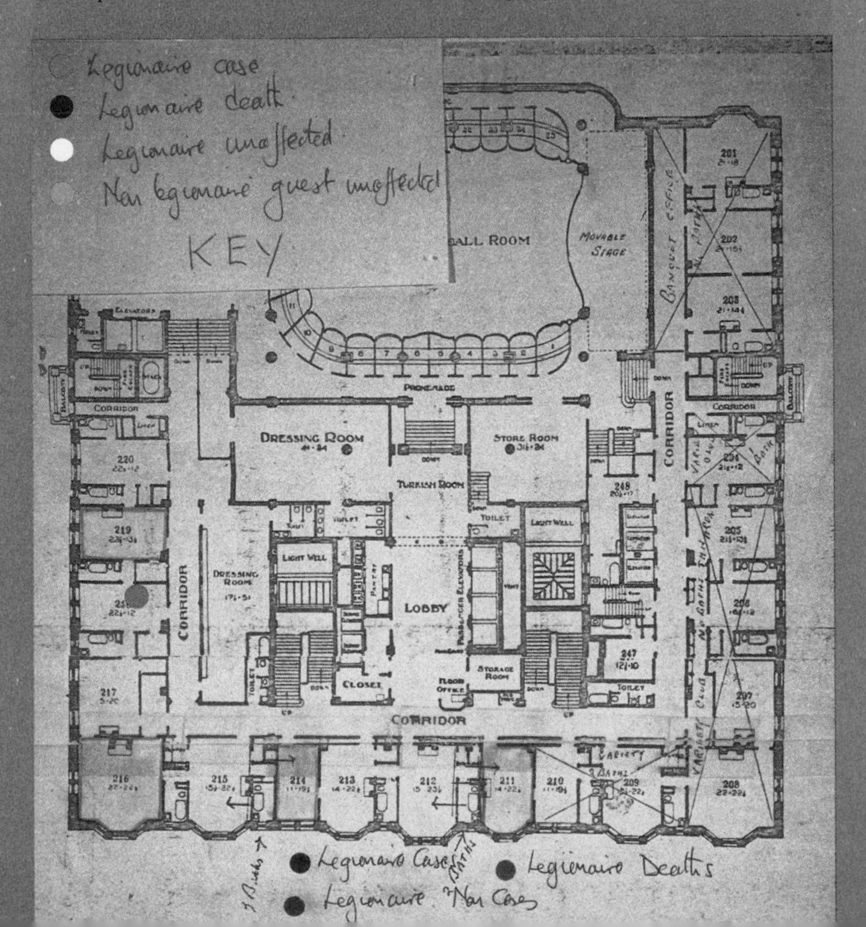

When a mysterious disease started killing healthy young people of the Navajo Nation in the Four Corners area of the southwestern United States, an army of medical detectives from state and federal agencies swung into action. In a remarkably short time, the pathogen that causes the disease and the small mammal that carries it were identified, through a combination of painstaking legwork, top-notch epidemiological methods, and advanced molecular biological techniques. What these medical scientists discovered meshed in many of its details with knowledge deeply imbedded in Navajo oral traditions.

Four Corners is the name given to the place where the borders of New Mexico, Arizona, Colorado, and Utah meet. The Navajo Indian reservation covers much of this normally arid region, and the Navajo people have lived in its high desert for centuries.

In May 1993, a young Navajo was on his way to the funeral of his wife, who had died five days earlier, when he began to have trouble breathing. He was taken to a rural hospital run by the Indian Health Service, and died there within a few hours. Autopsies were performed on the two young people by a pathologist from the New Mexico State Office of Medical Investigation, who noted similarities with another autopsy she had performed a month earlier. It wasn't long before two other sudden, unexplained deaths from respiratory failure were brought to the attention of hospital authorities, who reported a possible epidemic of an as-yet unknown infectious disease to the New Mexico Department of Health.

Within days, the departments of health of New Mexico, Arizona, Colorado, and Utah had been alerted, and shortly thereafter the federal CDC and the U.S. Army Medical Research Institute of Infectious Diseases (USAMRIID) were involved as well.

Lab tests quickly ruled out various possible causes, including bubonic plague (a now-rare rodent-borne illness that is occasionally seen in the southwestern United States), influenza, and herbicide poisoning. Tissue samples were sent to the special pathogens branch of the CDC in Atlanta, Georgia, where high-tech genetic tests were performed. One of these is called polymerase chain reaction (PCR), a method for rapidly synthesizing many identical copies of a DNA segment so that it can be more readily analyzed. Another test is called enzyme-linked immunoabsorbent assay (ELISA), which identifies the antibodies produced in response to a particular virus, and is often used to diagnose HIV infection as well. USAMRIID compared the genetic material with samples of genetic material from the U.S. Department of Defense's huge library of existing viruses.

It was quickly determined that the pathogen belonged to the family of hantaviruses, but was of a type previously unknown. By November, the virus had been isolated and dubbed Muerta Canyon Virus, later changed to Sin Nombre (no-name) virus.

The hantavirus is named for the Hantaan River in Korea, where the disease, which is carried by Asian striped field mice, is endemic. To date, fourteen different hantaviruses have been identified, and they generally fall into two groups: old- and new-world types. The six known old-world hantaviruses cause what is called hemorrhagic fever with renal syndrome (HFRS), which as the name suggests is characterized by internal bleeding and kidney failure. Of the eight new-world hantaviruses, four cause what is now called hantavirus pulmonary syndrome (HPS), characterized by the rapid onset of severe flu-like symptoms—fever, muscle aches, headache, and cough—leading quickly to respiratory failure and, in about half of all cases, death.

Because hantaviruses are known to be carried by rodents, a massive operation was launched to catch mice, rats, and other rodents in the Four Corners area. Between June and August of 1993, 1,700 rodents had been trapped inside and outside the homes of victims as well as in the places they worked and the groves of piñon trees in the area. (Piñons are evergreens native to the southwest; their nuts are the main food for the rodents living there.)

When the rodents were dissected, it became clear that the deer mouse was the carrier. Hantavirus was detected in thirty percent of the trapped deer mice, and in only a few of the other rodents.

One question still perplexed the disease detectives: Why the sudden outbreak? It turned out that after several years of drought, 1992 and 1993 saw an unusual amount of snow and, in the spring, to the melting snow was added greater than normal rainfall. The change in conditions was caused by the El Niño weather system. As a result, the piñons produced a bumper crop of nuts, which provided ideal conditions for a population explosion among the deer mice which thrive on them. The more deer mice, the more contact they had with people living in the area.

Hantaviruses are not transmitted through person-to-person contact, nor through the bite of an insect. Rather, the mouse carrier, which remains healthy, sheds the virus in its urine and feces, which are deposited wherever it happens to be. When dust containing the dried droppings is disturbed—when someone sweeps out a dusty cabin or tool shed, for example—the virus is inhaled along with the dust. The virus can also be picked up through the eyes or broken skin; when water or food contaminated with mouse excretions is ingested; or more directly, if a human is bitten by an infected mouse.

We now know that Sin Nombre virus is not a new disease, only a newly identified one. Analysis of lung tissues from people who died of unexplained respiratory diseases going back several decades has turned up evidence of the virus as early as 1959. And other types of hantavirus have been found in other areas of the United States: In Louisiana, the Bayou virus has been linked to the rice rat; in Florida, the cotton rat has been found to carry Black Creek Canal virus; in New York, the white-footed mouse is blamed for a virus called New York-1. Rodent-borne hantaviruses have also been found in Canada and the South American countries of Argentina, Brazil, Chile, Paraguay, and Uruguay. Nonetheless, it remains a rare disease, with fewer than 100 cases in 20 of the United States in the years since 1993.

The speed with which the Four Corners case was solved contrasts markedly with the time it took to isolate and identify the first hantavirus in Hantaan, Korea: six months versus several years. According to Joshua Lederberg, Professor Emeritus at The Rockefeller University, "The discovery . . . is one of the success stories of the integration of medical care, public-health oversight, and molecular diagnostic techniques." But while the scientific community was congratulating itself on the speed and efficiency with which the disease was identified and contained, the people of the Navajo Nation knew the natural history of the disease all too well. Within the memory of tribal elders there had been two previous outbreaks, in 1918 and 1933, two other years when, according to Navajo oral tradition, there had been more rain than usual and an abundance of pine nuts and the mice that eat them. The Navajo know that mice bring disease and should, therefore, be avoided.

In the Navajo belief system, disease comes when the balance of the universe has been disturbed, an occurrence that accurately describes the environmental changes brought about by El Niño, in many parts of the world, including Four Corners, U.S.A.

Scenes from the Southwest Four Corners region where the first U.S. hantavirus outbreak occurred in 1993.

Prairie deer mouse (*Peromyscus maniculatus*), the rodent reservoir host of hantavirus.

Altered weather conditions produced a large crop of piñon nuts in 1993, which allowed the deer mouse populations to explode.

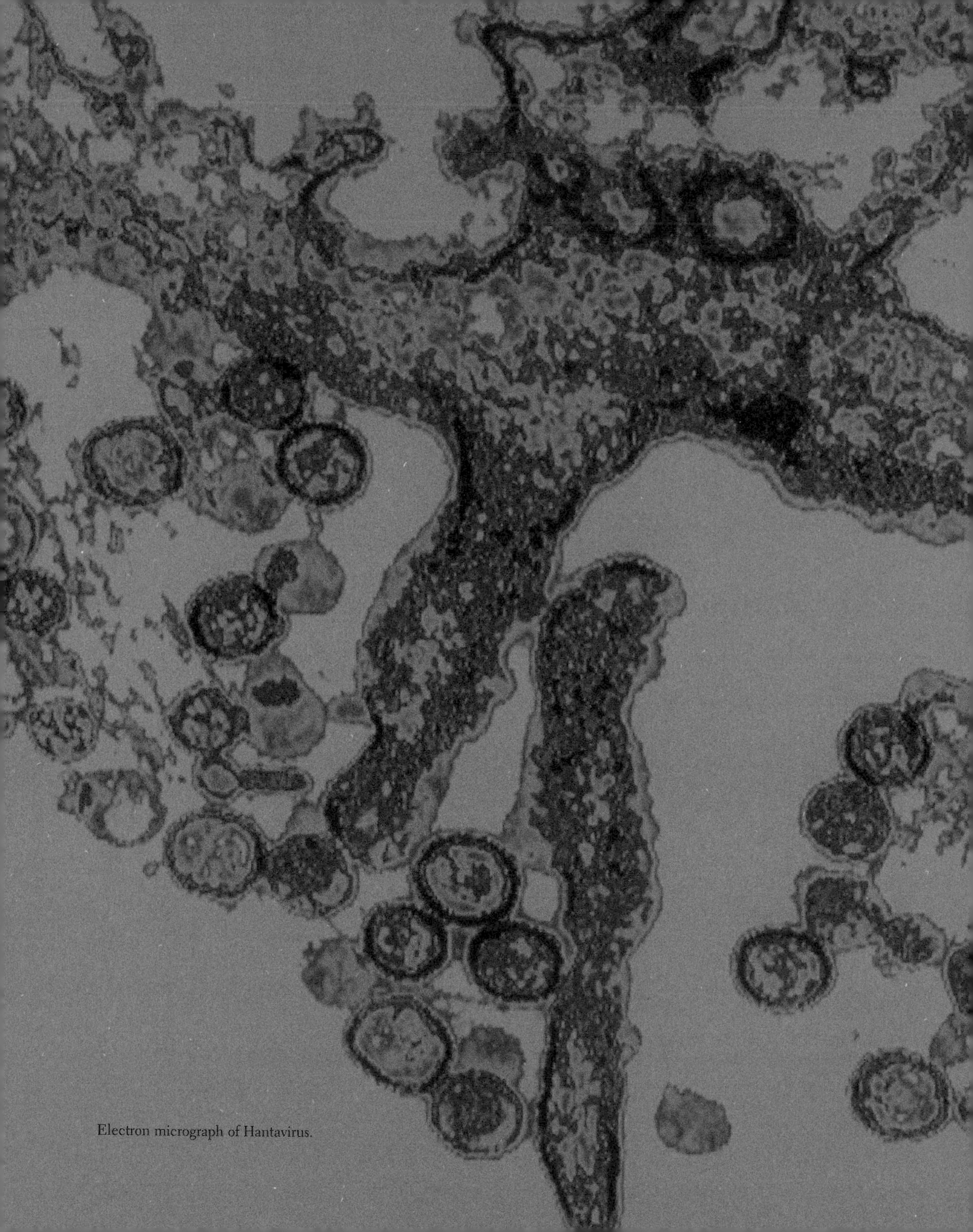

Electron micrograph of Hantavirus.

Using Remotely Sensed Data to Anticipate Risk Areas for Hantavirus Pulmonary Syndrome

JONATHAN A. PATZ AND GREGORY E. GLASS

INTRODUCTION

In 1997, we conducted a study to evaluate how satellite imagery could predict areas at high risk for hantavirus. The study focused on an analysis of the 1993 hantavirus outbreak in the Four Corners region of the United States and the satellite images taken of this area. The study was based on the hypothesis that the outbreak was caused by heavy rains of the strong El Niño of 1991–92. By surveying satellite images of the area, we hoped to prove the connection between El Niño and areas at potential risk for hantavirus so that proper health measures could be taken to contain a future outbreak.

BACKGROUND

During the spring and summer of 1993, the American Southwest was the scene of an outbreak of acute respiratory distress syndrome. Although the people contracting this disease had previously been in good health, more than half of them died, typically within a few days of their first symptoms. Brilliant and expeditious medical sleuthing soon established that these unfortunates were the victims of a previously unrecognized disease, now called the hantavirus pulmonary syndrome.

This newly discovered hantavirus, the Sin Nombre virus, occurs among many, but not all, populations of a common American field rodent (*Peromyscus maniculatus*), commonly known as the deer mouse. The virus maintains itself in these host populations—passing from mouse to mouse and

Jonathan A. Patz is a Research Assistant Professor and Director of the Program on Health Effects of Global Environmental Change at the Johns Hopkins School of Hygiene and Public Health.

Gregory E. Glass is an Associate Professor in the Department of Molecular Microbiology and Immunology at the Johns Hopkins School of Hygiene and Public Health.

from generation to generation—apparently without harming them. However, humans who come into direct contact with these infected mice or—more commonly—with their urine, feces, or saliva, can suffer dire consequences. There is no evidence of people catching hantavirus pulmonary syndrome from other people. Some evidence suggests that cases of hantavirus pulmonary syndrome had occurred long before this outbreak. In these cases, physicians had attributed the symptoms to other diseases. It was the essentially simultaneous clustering of the 1993 cases—the hallmark of an outbreak—that focused epidemiologists' attention and led to the recognition of hantavirus pulmonary syndrome and of the Sin Nombre virus.

Why did this outbreak of the hantavirus pulmonary syndrome occur when and where it did? Could this outbreak have been predicted?

THE HYPOTHESIS

According to one hypothesis, the hantavirus pulmonary syndrome outbreak of 1993 was an indirect consequence of the strong El Niño Southern Oscillation (ENSO) of 1991–92. During a southern oscillation, the trade winds that normally blow from east to west along the Equator become weaker, or even blow from west to east. When trade winds blow in their normal direction, they cause a huge mass of warm surface water—thousands of miles in area—to accumulate off the eastern side of the archipelago nation of Indonesia. However, when the trade winds blow from west to east, this huge mass of warm ocean surface water accumulates instead on the other side of the Pacific, off the coast of Peru. When that happens, an El Niño event occurs. Spanish for The Christ Child, the term El Niño was first applied, by Peruvian fishermen, to an annual warm southward ocean current that occurs off Peru during the Christmas season.

El Niño events occur sporadically, about every three to four years on average. During a mild event, the surface temperature of these vast warm regions of water may be 2 to 3 degrees Celsius (4 to 5 degrees Fahrenheit) above normal; during a strong event, the surface temperature may be 8 to 10 degrees Celsius (14 to 18 degrees Fahrenheit) above normal. During a strong El Niño event, the Earth's climate is temporarily, but strikingly, altered. The massive amount of evaporation from this warm mass of water occurs near Peru instead of near Indonesia. Consequently, the subsequent condensation and precipitation of this water occurs mainly in the western equatorial pacific region instead of in the eastern equatorial pacific region. Australia and Indonesia suffer droughts while rainfall greatly increases in many regions of the Americas, even causing massive flooding in desert regions.

A strong El Niño event began in 1990 and continued until mid-1995. This El Niño may have been responsible for the intense rains that occurred in the Southwest United States in 1991 and 1992 during the usually dry spring and summer months. It may also have caused that area's mild winter in 1992. This unusual weather, in turn, may have led to significant increases in rodent populations in local areas scattered throughout the region. It would have done so by increasing the growth of vegetation that provides food and shelter to rodents. The extra growth of vegetation may have caused the local increases in rodent populations which in turn may have triggered the hantavirus pulmonary syndrome outbreak.

Several lines of evidence support this hypothesis. First, an ENSO significantly affects weather—rainfall amounts and temperature, in particular—in the Southwest. Therefore, it is likely to influence plant growth in local environments. Second, limited monitoring showed that some rodent populations increased dramatically there—in some cases by factors of ten to fifteen—during this ENSO. Third, studies showed that rodent abundance varied significantly even between regions separated by short distances, and that rodents were especially abundant where hantavirus pulmonary syndrome occurred. However, directly connecting the 1991–92 ENSO to the hantavirus pulmonary syndrome outbreak requires showing that areas with especially large amounts of plant growth were more likely to be the scenes of hantavirus pulmonary syndrome.

Establishing the connection between this outbreak and ENSO might make it possible in the future to identify regions at increased risk for hantavirus pulmonary syndrome. These would be regions in which the local flora was growing in unusual abundance. Identifying these regions sufficiently far in advance could galvanize public health officials and health care providers into action. They could inform the public of the increased risk of this infectious disease and suggest how to reduce the risk of being exposed to it. This would also allow physicians to be more prepared in diagnosing the disease, thereby reducing fatalities from the outbreak.

TESTING THE HYPOTHESIS: REMOTE MONITORING

Electronic images of the Earth captured by an orbiting satellite and then radioed to stations on Earth can rapidly provide information about ground cover, weather patterns, or the effects of weather on vegetation. These images result from electromagnetic radiation that strikes sensors after it has been reflected or transmitted from the target area of the Earth's surface. For example, the orbiting LANDSAT Thematic Mapper satellite platform receives electromagnetic energy from the surface in seven bands (regions) of the electromagnetic spectrum. Three of these regions are within the visible light portion, and four are within the infrared (or radiant heat) portion. The light received in the seven bands is the primary data used to identify the surface features on that part of the Earth's surface emitting or reflecting the radiation.

Computers rapidly "correct" this data to take into account the position of the sun as it is illuminating the observed portion of the Earth's surface. This would affect the intensity of the sunlight illuminating the surface. Other corrections involve the Earth's topography at the observed location. Correction for topography is needed because the slope of the surface and the direction the slope faces affect the intensity of the surface radiation reaching the satellite sensors.

Computer analysis of this data provides a "spectral pattern" or "spectral fingerprint" that corresponds to specific surface features. For example, green light arriving from the surface says vegetation—green plants absorb the blue and reddish portion of sunlight and reflect the green portion. The pattern of relative "brightnesses" among the three infrared bands shows temperature, among other surface characteristics.

TESTING THE HYPOTHESIS: CORRELATING 1992 SATELLITE IMAGERY WITH 1993 HANTAVIRUS PULMONARY SYNDROME RISK

Satellite images recorded in mid-June 1992 of a 105,200 square kilometer area of the Southwest were carefully aligned using U.S. Geographical Survey topographical maps. In this area, twenty-seven cases (ninety percent of all cases) of hantavirus pulmonary syndrome were identified between November 1992 and November 1994. The locations of the homes of the twenty-seven hantavirus pulmonary syndrome victims and of 170 randomly selected people—the controls—were carefully established using the U.S. Geographical Survey maps or measurements made using the Global Positioning System.

This data was analyzed using a type of statistical analysis called logistic regression analysis. The analysis used data gathered in June because this is before the summer monsoon season during which much of the region's rain occurs. This analysis showed a significant correlation between the hantavirus pulmonary syndrome cases and regions whose satellite images were unusually bright in three of the six spectral bands. The data showed that these bright regions had contained moist vegetation in June 1992, a year before the hantavirus pulmonary syndrome outbreak. The data also showed that hantavirus pulmonary syndrome risk increased at higher elevations. Green growing vegetation before the monsoon season is a biologically important hantavirus pulmonary syndrome risk factor in the Southwest.

DISCUSSION

This study shows that satellite imagery can identify local areas at high risk for hantavirus pulmonary syndrome sufficiently far in advance for effective public health interventions to occur. It can also rule out hantavirus pulmonary syndrome for the large portion of the population. This could help doctors reach a proper diagnosis for patients with symptoms that misleadingly suggest hantavirus pulmonary syndrome. In particular, the LANDSAT Thematic Mapper data revealed that local areas which were much more heavily vegetated than most rural, human-occupied sites in the area were at increased risk for hantavirus pulmonary syndrome. They were much more likely than average to be sites where a case of hantavirus pulmonary syndrome subsequently—a year or so later—occurred. This is consistent with these sites favoring abundant populations of deer mice—some harboring the Sin Nombre virus that causes hantavirus pulmonary syndrome. (Definitive proof of this will depend on the results of field surveys which will determine the population sizes of rodents, and the incidence of Sin Nombre virus infection within these populations, in areas satellite imagery shows to be at increased risk because heavily vegetated.) The results support previous observations that variations in environmental conditions over relatively short distances correlate with significant differences in rodent population densities. They also support epidemiological investigations that found that the only measurable risk factor around hantavirus pulmonary syndrome sites during the 1993 epidemic was the abundance of deer mice. (The type of housing occupied by people living in this region had no measurable relationship to their likelihood of contracting hantavirus pulmonary syndrome.)

In the future, field investigators can investigate the deer mouse population in regions that satellite imagery suggests may be at increased risk of hantavirus pulmonary syndrome to see if they harbor the Sin Nombre virus. Such information will allow public health resources to be targeted even more efficiently.

Analysis of satellite images collected in 1997—again, during another strong El Niño event—were used to predict areas of hantavirus pulmonary syndrome risk for 1998. For the most part, these sites were in different locations than those at risk in 1992. This suggests that regions at risk of hantavirus pulmonary syndrome vary from one year to the next. Perhaps this is because each small area of the Earth's surface does not get the same amount of rainfall from one year to the next—even during ENSO periods.

Generally, the analysis of satellite imagery provides a method to survey large geographic regions efficiently for environmental indicators of disease risk. This technique could make it practical from a public health standpoint to monitor the risk of rare diseases in which animals carry the infecting pathogens. For example, satellite imagery detected environmental conditions at various sites in East Africa that correlated with general levels of Rift Valley fever virus activity. Similar data has identified sites at risk for such diseases as Lyme disease (transmitted by ticks), fascioliasis (a liver disease caused by infecting flatworms), sleeping sickness (transmitted by tsetse flies), and schistosomiasis (a flatworm infection).

Many analyses suggest that global warming, currently occurring and expected to accelerate, will significantly alter the climate in many regions of the world. One aspect of this climate change will be more frequent and more severe droughts in some locations, and drastically increased rainfall in other locations. ENSO events consistently cause both severe droughts and excessive precipitation. Therefore, they can give us a preview of how future global warming might affect the patterns of diseases transmitted via vectors.

How does the process of satelite visualization help reconstruct the chain of events that led to hantavirus pulmonary syndrome?

How could satellite visualizations be used for tracking changes in the environment, population, urbanization, and how would these factors add to an understanding of infectious disease?

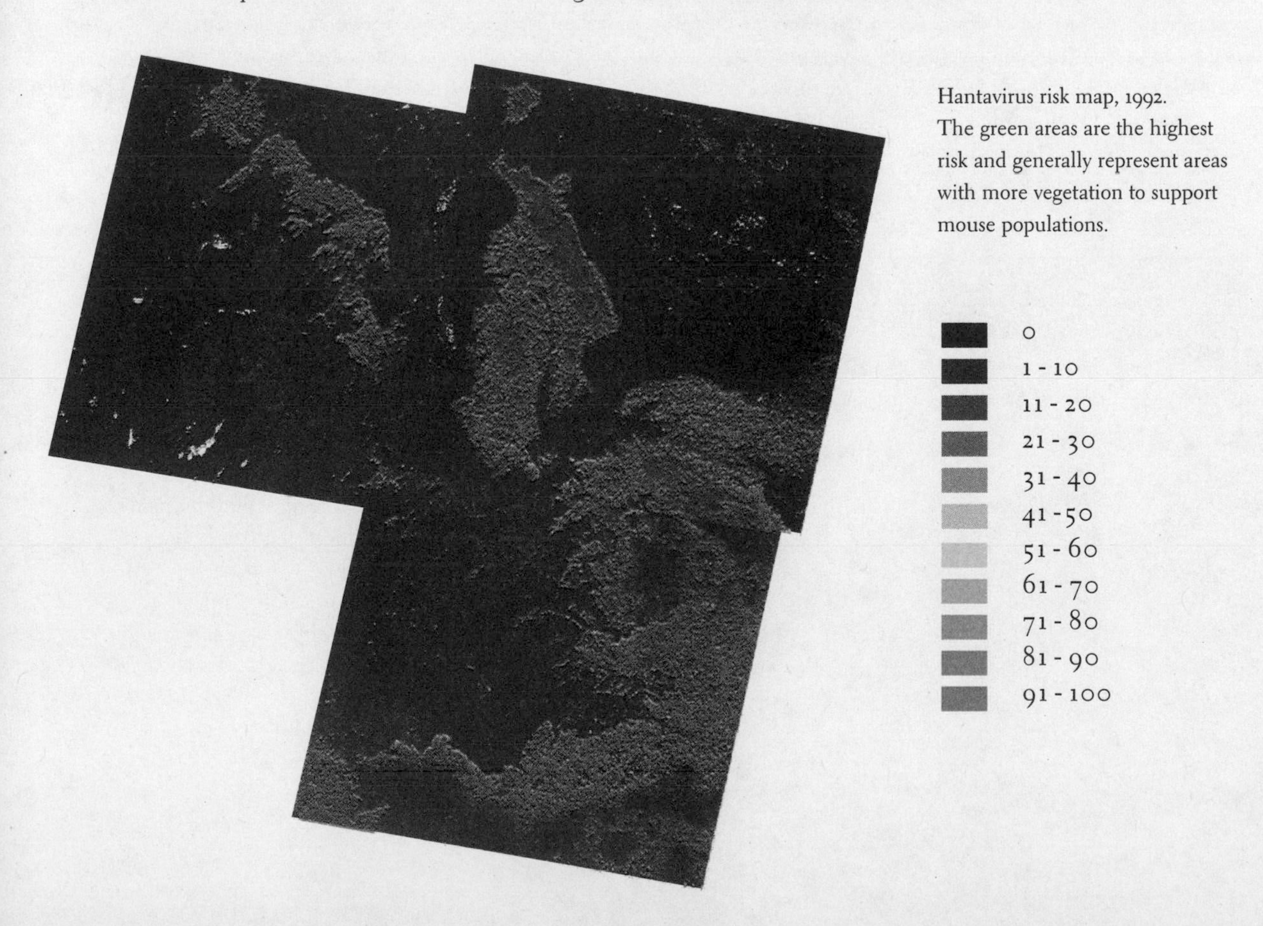

Hantavirus risk map, 1992.
The green areas are the highest risk and generally represent areas with more vegetation to support mouse populations.

Case Study

The Two-Way Race to Find the AIDS Virus

The story of the discovery of the virus that causes AIDS is also the story of a clash of powerful personalities, a rivalry so fierce that it threatened to derail research and took the efforts of diplomats from two of the most powerful nations in the world to quell the fray. Two highly regarded researchers, the American Robert Gallo and the Frenchman Luc Montagnier, have both claimed credit for isolating the virus that destroys the immune system and leads to the disease we call acquired immune deficiency syndrome, or AIDS.

The human immune deficiency virus (HIV) belongs to a class of pathogens referred to as retroviruses. Medical science has known about retroviruses for a long time, but until 1980, they were thought not to infect humans. Robert Gallo led the team that discovered the first retrovirus responsible for a human disease. It was given the name HTLV-1, for human T-lymphotropic virus type 1, and it causes the blood cancer adult T-cell leukemia (ATL). Since then, a second cancer-causing virus, HTLV-II, was discovered. Both viruses are transmitted through infected blood, sexual intercourse, and from mother to child during pregnancy.

In 1981, while the study of human retroviruses was in its infancy, disturbing reports started coming in to the Centers for Disease Control and Prevention (CDC), the federal government's center for epidemiology, based in Atlanta, Georgia. The first were five cases of a very rare pneumonia caused by the protozoan *Pneumocystis carinii* (PCP), all of which had been seen in Los Angeles within just a few months. Around that same time, the CDC also received reports of a surge in another rare disease: twenty-six cases of Kaposi's sarcoma (KS) in thirty months. KS is a cancer usually seen only in elderly men, but the victims in these cases were young men and, like those who had come down with PCP, they were all homosexuals.

Soon news spread throughout the medical community, and shortly thereafter through the country as a whole, about a "gay epidemic." We now know that these men were suffering from AIDS, and that KS and PCP were signs that their immune systems had been destroyed. In the two decades since, both the disease, which has grown to worldwide epidemic proportions, and our knowledge of it have increased. We now know that the disease is not limited to gay men. Intravenous drug users and their sexual partners, hemophiliacs and others receiving HIV-contaminated blood as part of a medical procedure, newborns of HIV-infected mothers, and people who have unprotected sex with infected individuals are also at risk. Throughout the world, forty-seven million people have been infected with HIV and fourteen million have died. An arsenal of drugs has been developed to control the infection and the many other conditions that attack people whose immune systems have been damaged by it. Public education efforts aimed at preventing transmission go hand in hand with research to find a cure. But none of this knowledge could have been acquired if the microbe responsible had not

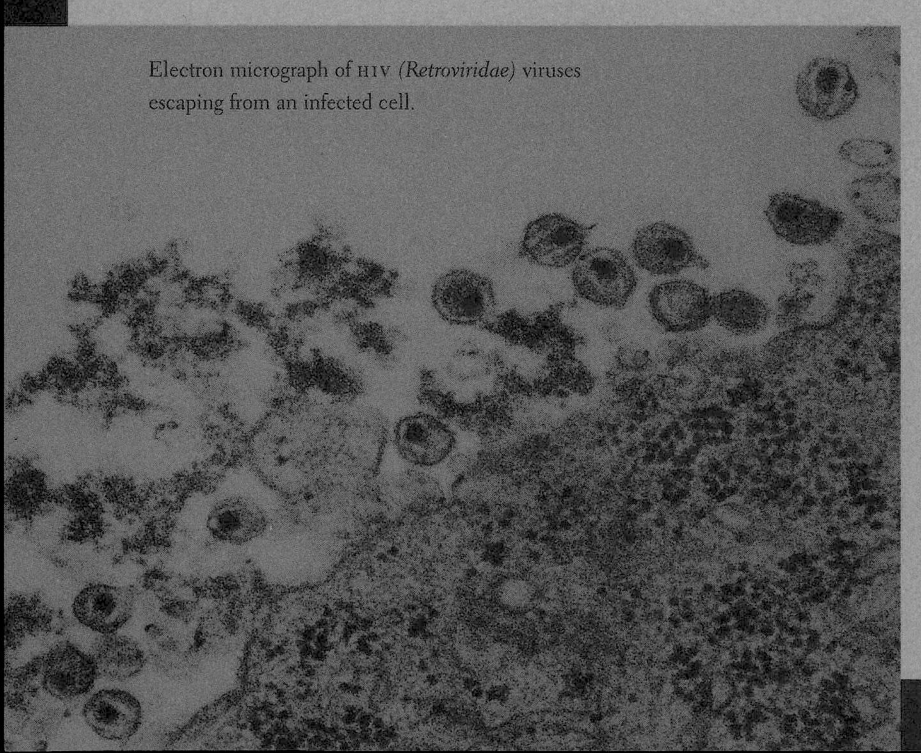

Electron micrograph of HIV *(Retroviridae)* viruses escaping from an infected cell.

been isolated and characterized.

Gallo suspected a retrovirus, based on evidence that, like the HTLVs, it seemed to spread through blood and sex, and attacked the T cells (vital parts of the immune system) with the result that infected people were unusually vulnerable to other infections that those with healthy immune systems easily combat.

At about the same time, researchers in France were also trying to identify the infectious agent that was causing the new disease. They found it was a retrovirus, but when they tested it to see if it was one of the HTLVs, the tests were negative.

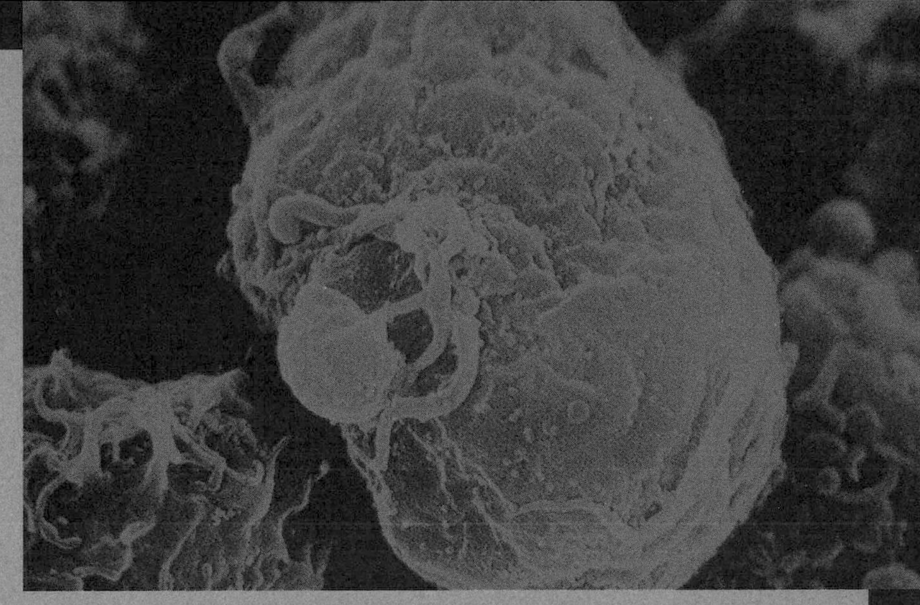

Scanning electron micrograph of a T cell being infected by HIV *(Retroviridae)*.

The French researchers decided it was a new virus and named it LAV (for lymphadenopathy-associated virus, because it was taken from the swollen lymph node of a young man in the early stages of AIDS). In 1983, Luc Montagnier, head of the French team, declared that LAV was most likely the pathogen responsible for AIDS.

On the other side of the Atlantic, Gallo and his team were skeptical and continued to try to find a retrovirus that could be associated with the growing number of people turning up with the disease. By the end of the year, they came up with their own candidate, and named it HTLV-III.

Up to that point, the race to find the mysterious virus was a friendly rivalry and the two labs shared specimens and laboratory tests for characterizing viruses. But then a series of events occurred that shook the scientific community. First, it was found that LAV and HTLV-III were actually the same organism. In 1986, the International Committee on the Taxonomy of Viruses voted to call it HIV, human immune deficiency virus, and all agreed it was the cause of AIDS.

It was not long before an argument broke out about who had discovered the virus first. At stake was more than the prestige and possible Nobel Prize; it was also a matter of the patent for a test to detect the virus and the millions of dollars in royalties that would come to whomever held it. The French insisted they had made the discovery first and, moreover, that Gallo had stolen from Montagnier's laboratory the viral specimen on which he based his studies.

After first denying the charges of theft, Gallo later conceded that his viral samples may have been contaminated with specimens sent to him by Montagnier's lab. Proving that, he said, would be impossible, and should not be allowed to interfere with the importance of the discovery. Montagnier, for his part, replied that, at best, Gallo's lab had merely rediscovered the virus and for that deserved no credit and no part in the patent.

By 1987, the battle had gotten so far out of hand that a deal was worked out, and signed by President Ronald Reagan and French Prime Minister Jacques Chirac, allowing a fifty-fifty split in any royalties earned from the AIDS blood test based on the Gallo/Montagnier findings. Later, both countries agreed that the French patent holders should get a larger share.

Fortunately for the field of AIDS research and the people afflicted with the disease, however, everyone else has moved on. No one has discovered a cure for AIDS, but as we learn more about the natural history of the virus—how it is transmitted, how it infects human cells, and even more important, how its deadly work can be interrupted—there is hope that what was a few short years ago a death sentence will instead be a chronic (lasting), but manageable disease.

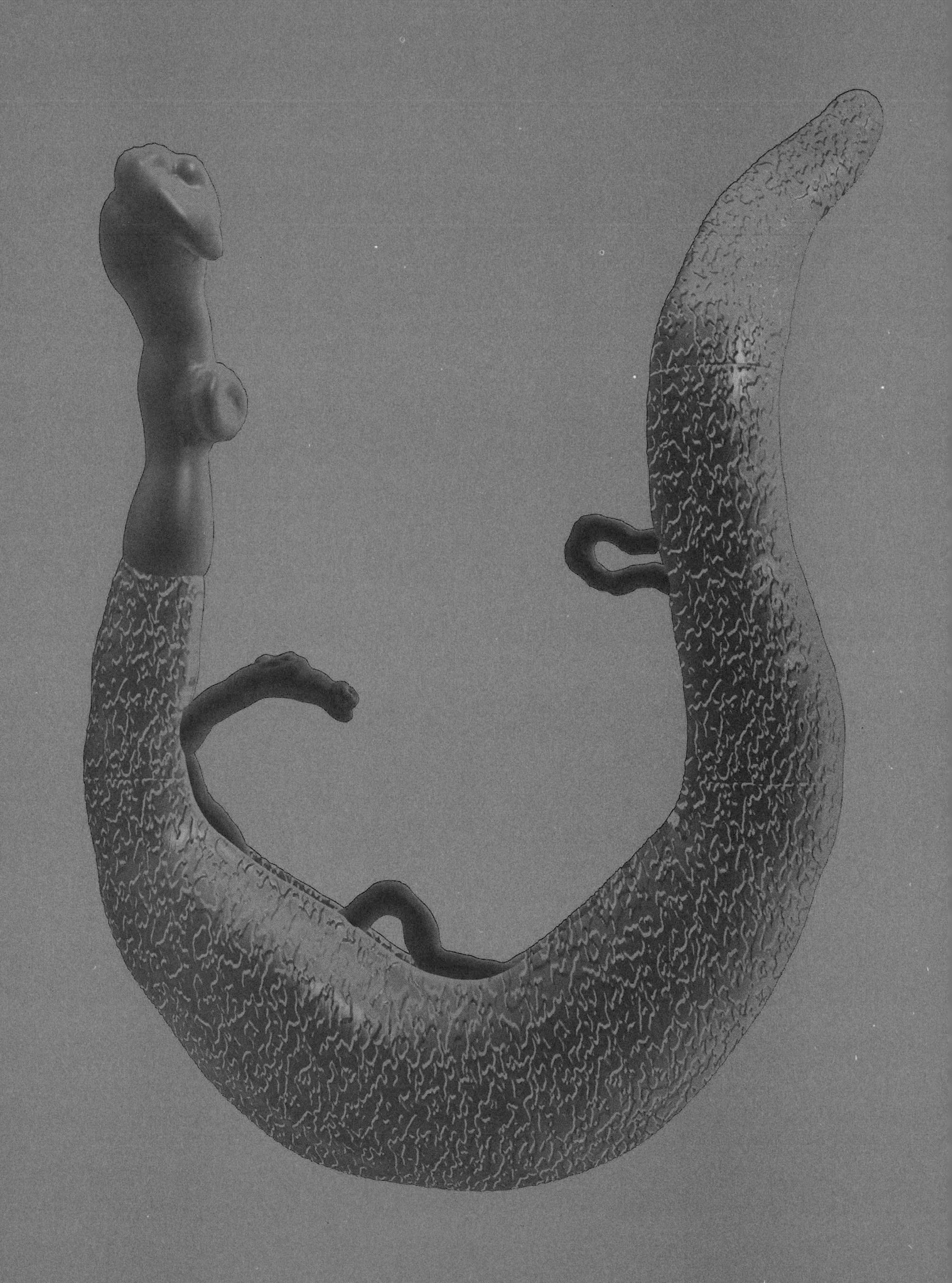

Section Five: Epidemics and Pandemics

left Computer-generated image of schistosomiasis-causing worm (*Schistosoma mansoni*).

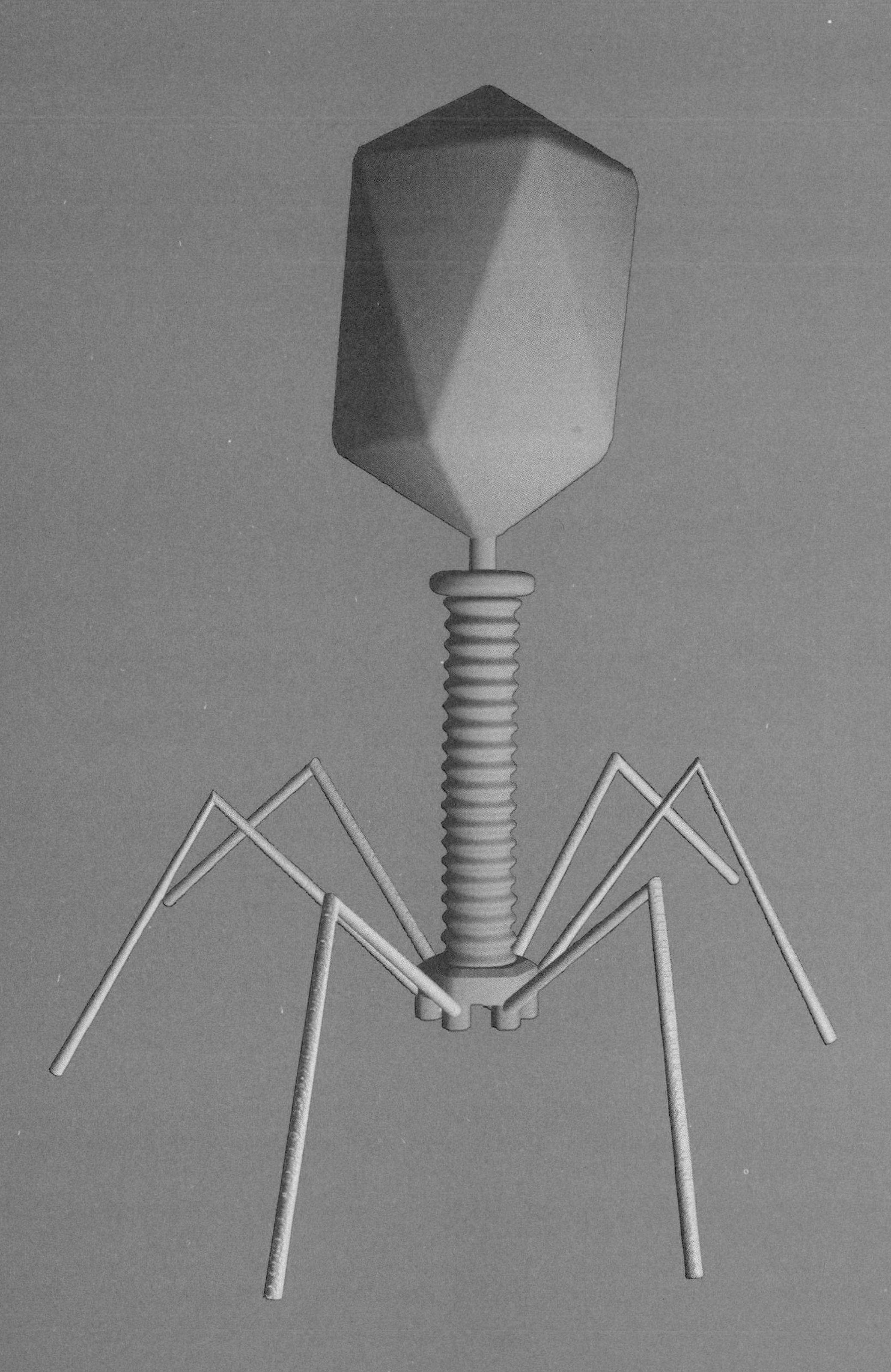

Computer-generated image of a bacterial virus (T-4 Bacteriophage).

Section Five: Epidemics and Pandemics

Introduction ROB DESALLE

The term epidemic refers to any disease, injury, or health-related event that occurs suddenly among more people in a particular region than is normal. Textbook definitions suggest that an epidemic is the "joining together of populations" because they share the same disease. From the point of view of an ecologist, a single population occupies a single ecosystem. Therefore, populations joined by an epidemic behave as a single ecological unit. An epidemic could act as a force of natural selection, perhaps killing people with certain genes but sparing people with others. In that sense, an epidemic also links separate populations into a single evolutionary unit. A pandemic is an epidemic that occurs simultaneously in many different parts of the world. From an ecological perspective, a pandemic temporarily connects many, perhaps even all, humans into a single ecological and evolutionary unit. To understand epidemics and pandemics, we need to expand the concentric ring to include interactions between populations that allow disease to spread over a large region—or even over the entire globe.

Two factors, in particular, are important in epidemics: (1) the patterns of, and reasons for, the movement of individual people, groups of people, or entire populations, from one area to another; (2) the modes through which the movement of people spreads disease.

There are four major factors that influence how infectious disease moves from one geographical area to another: war, trade and travel, urbanization, and global climate change.

War is one of the ways that infectious disease can travel. Most people think that once a war ends, the abnormally high incidence of death, disease and environment damage also ends. In reality, however, the tragedies of war often continue long after it has ended. When troops move into new regions, they encounter native peoples and native ecosystems. They bring in new diseases that infect the native population and they bring in alien species that may disrupt the ecosystem. When military personnel return home, they may bring new diseases back with them. These diseases can then infect their fellow citizens. Returning military personnel may also bring back alien species that may disrupt their ecosystem.

During World War I, ten million people died as the direct result of the use of weapons. The indirect devastation was vastly greater. More than twenty million people died in the 1918–21 influenza epidemic that spread to nearly all parts of the globe, including places not directly connected to the conflict in Europe. This flu epidemic spread largely because of the movements of infected military personnel or infected refugees. The extremely high number of fatalities resulted in large part from malnutrition, which lowered people's resistance to disease. This malnutrition was an indirect result of the ecological damage that occurred during the war. In particular, this ecological damage greatly reduced crop outputs, creating severe worldwide food shortages.

A second way that infectious disease can move from one place to another is through trade and travel. Our globe shrinks every day as people and goods travel ever more quickly. Insects, other vectors, and the infectious microbes themselves hitch rides on traveling people and goods—often causing outbreaks, or even epidemics, when they reach a new location. Historically, the migration of peoples has been a major cause of the spread of infectious disease. Europeans brought infectious diseases with them when they explored and colonized the New World and the South Pacific. The results were devastating to such native peoples as the Amerindians, the Maoris of New Zealand, the Australian Aborigines, and the Hawaiian Pacific Islanders. Having never been exposed to the European diseases carried by these travelers, these native peoples lacked resistance to the microbes causing these diseases. Consequently, they became infected—and died—in massive numbers.

Urbanization is a third way that infectious disease travels. An overriding tendency among modern humans has been to abandon a rural farming life and move to cities. In 1900, twenty percent of all people lived in cities; today more than eighty-five percent of all people live in cities. In 1900, only five cities had populations larger than one million; by the year 2020, there will be twenty-five megacities worldwide, each with more than twenty million people. When moving to cities, people abandon the ecosystems in which their various biological adaptations arose through evolution. They also create new ecosystems in an urban setting in which they are not so well adapted biologically. Moreover, the crowded conditions in urban regions favor the transmission, both directly and indirectly, of pathogenic microbes from person to person. Overall, the many new opportunities urbanization has created for microbes have been a major factor in the spread of disease across the globe.

The fourth way that infectious disease can move from one place to another is through global climate change. Human activity can, and has, greatly altered the environment. Among these effects may be a significant increase in the greenhouse effect. Mainly through deforestation and the combustion of fossil fuels, humans have significantly increased the concentration of carbon dioxide in the Earth's atmosphere. This in turn increases the proportion of infrared radiation emitted from the Earth's surface that changes into heat in the lower atmosphere, instead of escaping to outer space. If present trends continue, the Earth's average surface temperature may increase by several degrees Fahrenheit by the year 2100. Such an increase could have devastating effects on the spread of such tropical diseases as malaria. Today, malaria typically does not occur outside the tropics because the insect vector that carries the microbes cannot survive cold winters. However, global climate change will significantly increase the range in which the mosquito species that are vectors for malaria can live and breed. Malaria will then occur and persist in regions where it never occurred or was once eliminated. This human-derived global climate change may also alter the frequency and intensity of the El Niño events. Such climate changes could alter precipitation patterns, which could alter vegetation patterns and, in turn, alter the distribution of animal species that are vectors for a variety of infectious diseases.

Today, humans are part of a "global village." They can receive information from, and send information to, nearly any part of the world almost instantaneously. Moreover, people, vectors, and microbes, as well as medicines and medical information, can travel around the globe with great frequency and ease. To understand how infectious disease emerges, spreads, and is treated in our global village, we must understand both the global ecology and the differing cultural perspectives on infectious disease. Three points are key here:

- Different cultures perceive disease differently. They differ in their perceptions of what disease is, what causes a specific disease, how it spreads, and whether or how to control it. It is necessary to consider these differences to understand the full spectrum of difficulties that must be overcome to combat disease on a global scale.
- Even within a single country, different socioeconomic, cultural, and ethnic groups are affected differently by, and respond differently to, a particular disease. Successful control of that disease requires understanding these differences. Different methods of disease control may be required, depending on the group in question.
- The epidemiology of a particular disease—including its prevalence and how it spreads—may differ from one culture to another. For example, in sub-Saharan Africa, the HIV virus that causes AIDS spreads mainly through heterosexual contact. In contrast, in North America and Europe, AIDS spreads mainly through injection of intravenous drugs and through homosexual contact.

Understanding these differences is a prerequisite for understanding the natural history of an infectious disease on a global scale. Particularly relevant are the differing public health strategies adopted by different countries and cultures. Such differences include whether or how to educate the public about particular infectious diseases, and about how to control the disease. If one country or culture wishes to persuade another country or culture to adopt specific preventive methods against a particular disease, it must provide its advice in a form that is palatable to the intended recipient. Otherwise, what may appear as sound advice on the sending end, might appear as cultural imperialism on the receiving end.

To explore epidemics and pandemics, I pose the following questions:

How dangerous are epidemics and pandemics?
In his essay, A. David Brandling-Bennett, Deputy Director, Pan American Health Organization, describes major historical epidemics to show how disease can change the course of history.

How has modern life contributed to the resurgence of infectious disease?
ANNE PLATT MCGINN, Research Associate, World Watch Institute, documents the major ecological and social changes globally that have led to recent epidemics of diseases we once thought were eradicated.

How should governments respond to outbreaks?
MARGARET HAMBURG, Assistant Secretary for Planning and Evaluation, United States Department of Health and Human Services, chronicles the response of New York City to tuberculosis to demonstrate critical issues in formulating public health policy.

How is the tragedy of AIDS in Africa a classic example of epidemic and pandemic occurence?
JOHN CALDWELL, Professor Emeritus of Demography and Coordinator of the Health Transition Centre, Australian National University, along with PAT CALDWELL, also of the Health Transition Centre, Australian National University, outline the ecological, social, and political factors contributing to the AIDS epidemic that is ravaging sub-Saharan Africa.

By placing epidemics and pandemics in historical and cultural contexts, the essays in this section address the complex relationship humans have with pathogens and how modern societies have changed the course of infectious disease.

At work in the lab.

Gel of Polymerase Chain Reaction (PCR).

Our Long Struggle Against Epidemics

A. DAVID BRANDLING-BENNETT

Although epidemics have visited human populations for millennia, the historical record of these visits is frustratingly incomplete. Ancient writers had an extremely limited understanding of the causes and nature of infectious disease; consequently, even when we have their accounts of epidemics, they frequently lack the information we need to identify the diseases they were describing. Nevertheless, we have clear evidence—both historical and physical—for the antiquity of smallpox. Three Egyptian mummies from the eighteenth and twentieth dynasties (1570 to 1085 B.C.)—including the mummy of pharaoh Rameses V—have skin lesions characteristic of smallpox; contemporary historical accounts suggest that smallpox struck the Hittite Empire (in today's Turkey and northern Syria) after its war with the Egyptian Empire in the thirteenth century B.C. Indian medical texts written in Sanskrit before 400 A.D. describe a disease very much like smallpox that probably occurred as early as 1500 B.C.; contemporary accounts suggest that Alexander the Great's army suffered from smallpox while in India in 327 B.C. In China, ancient texts refer to a disastrous epidemic, apparently of smallpox, that arrived from the north and spread throughout the empire in 243 B.C. Smallpox continued to take a heavy toll throughout history until its eradication in 1977.

Another epidemic disease reported by early historians is the plague. The first major epidemic of plague to affect Europe spread throughout the Roman Empire in 541 A.D. The most serious plague epidemic, called the Black Death, killed up to one third of the population (about twenty-five million people) of fourteenth-century Europe, starting in 1346. The plague finally abated in Europe in the seventeenth century, but only after the Great Plague of London killed an estimated 70,000 Londoners, about fifteen percent of the city's population, from 1664 to 1665. As recently as 1894, the plague killed from 80,000 to 100,000 people in Canton and Hong Kong, before spreading to other

A. David Brandling-Bennett is Deputy Director of the Pan American Health Organization.

countries; during the next twenty years, more than ten million people worldwide died in this epidemic.

After the fifteenth century discovery of the New World, the Spanish appear to have introduced smallpox, measles, and other epidemic diseases to Central and South America, decimating the indigenous populations. For example, the Aztec population in Mexico decreased from about thirty million people to 1.5 million during the first fifty years after the arrival of the Spanish, primarily from infectious diseases.

Although epidemics have long afflicted the human population, the ancient Greeks appear to have been the first to recognize epidemics as noteworthy phenomena, and to attempt to distinguish epidemics from other patterns of disease. Hippocrates (ca. 460 B.C. to ca. 377 B.C.)—of Hippocratic oath fame—applied the term epidemic to diseases that visit a community; he applied the term endemic to diseases that reside within the community. Although Hippocrates and later writers understood that epidemics were the unusual occurrence of diseases in populations, they were unable to agree on what caused them. Diseases overall were variously attributed to divine intervention, to an imbalance of the four humors (phlegm, blood, black bile, and yellow bile) in the body, to an innate flaw in the affected persons, and to changes in the atmosphere or in other aspects of the environment.

In 1546, the Italian physician and poet Girolamo Fracastoro proposed that smallpox and measles spread through the passing of a rapidly multiplying disease-specific seed from an infected person to a non-infected person. In 1675, the pioneering Dutch microscopist Anthony van Leeuwenhoek discovered that stagnant water contained small living organisms called protozoans which he termed animalcules. In the eighteenth century, many believed that animalcules caused such diseases as smallpox. However, rival explanations for infectious diseases still were prevalent, even dominant.

The germ theory of disease gained general acceptance only in the latter half of the nineteenth century. In the mid-1830s, the pioneering Italian bacteriologist Agostino Bassi concluded from twenty-five years of research that "silkworm" disease was contagious and that its cause was a microscopic fungus that spread by contact. In 1835, he proposed that many other diseases—including human diseases—were caused by parasitic plants or animals. Similarly, the German pathologist Friedrich Gustav Jacob Henle argued in 1840 that the causes of contagious diseases were parasitic microorganisms. In the 1850s, the famous French scientist Louis Pasteur carried out a brilliant series of experiments proving that living microorganisms caused the fermentation of wine and the souring of milk; in the 1860s, he proved that bacteria did not arise spontaneously from inorganic matter, but came from living (parent) bacteria. Finally, in the 1870s and 1880s, the German physician Robert Koch, a student of Henle, provided compelling support for the germ theory of disease: he proved that anthrax (in cattle), tuberculosis, and cholera resulted from infection by specific bacteria. In experiments that included culturing the pathogenic bacteria outside an animal's body and then causing the particular disease by inoculating a healthy animal with the cultured bacteria, Koch established principles for proving that a specific microorganism caused a specific disease.

One of the first to apply the germ theory of disease to the solution of a severe public health problem was the physician John Snow, who used the theory to explain the occurrence of cholera in London in 1854. In a piece of brilliant detective work, Snow associated a cholera epidemic with water from the Broad Street pump in Soho's Golden Square; he stopped the epidemic by removing the handle from the pump. He subsequently linked cholera to drinking water from the Thames River downstream from London—and therefore where contamination with the city's sewage was more likely. Snow's seminal work, the first scientific study of epidemics, represents the beginning of modern epidemiology.

John Snow was not alone in his concern about epidemics. Infectious disease was so important in the nineteenth century that it stimulated the first International Sanitary Conference held in Paris in 1851; the conference proposed international health measures to prevent the movement of epidemics from one country to another. In 1902, the countries of the Americas met in Washington, D.C., to establish what would eventually become in 1949 the Pan American Health Organization; its goal was to prevent the spread of infectious diseases such as cholera, smallpox, yellow fever, and plague through international commerce.

The discoveries of the last 150 years have led to the development of antimicrobial drugs and vaccines, which have been highly successful in treating and preventing infectious diseases. Even so, infectious disease remains the world's leading cause of premature death, killing more than seventeen million people each year, nine million of them children. That means 50,000 preventable deaths each day. Tuberculosis takes the highest toll of any infectious disease, killing more than three million persons each year, mostly adults. Acute respiratory infections, diarrheal diseases, and malaria each cause two to four million deaths, mostly in children. Although the poorest countries have the highest rates of illness and death from infectious disease, all countries continue to experience epidemics of various types and causes. The size of an epidemic may vary from a few cases of an unusual or previously controlled disease to thousands of cases of a disease that spreads widely through a community.

When a disease affects many countries, we may call it a pandemic. During the nineteenth century, five pandemics of cholera occurred, with the disease spreading from the Indian subcontinent to many countries throughout the world. Two pandemics of cholera have occurred in the twentieth century, the current or seventh pandemic of El Tor cholera having begun in Indonesia in 1961. It moved from there throughout Asia and Oceania and reached Africa in 1970, spreading rapidly throughout that continent over the next three years. All the countries of the Americas had remained free of epidemic cholera through the twentieth century until 1991, when the disease struck with force in Peru and then infected fourteen other countries in the Western Hemisphere in the same year. From 1991 to 1997, cholera afflicted more than 1.2 million people and caused nearly 12,000 deaths in the Americas.

Some epidemics are caused by parasites, bacteria, or viruses that researchers have discovered during the last quarter of the twentieth century. *Cryptosporidium parvum*, a parasitic protozoan that produces acute and chronic diarrhea, caused an outbreak affecting more than 400,000 persons, more than fifty fatally, in Milwaukee, Wisconsin, in 1993. An epidemic of pneumonia among Legionnaires, members of an American legion attending a convention in Philadelphia in 1977 (182 Legionnaires contracted the disease, and twenty-nine died of it) led to the discovery of the bacterium *Legionella pneumophila*. One of the most infamous new agents is the Ebola virus, which investigators discovered in 1976 and has produced epidemics in Sudan, Zaire, and other African countries. Several popular books and movies have capitalized on public concern that this virus, which is fatal in up to eighty percent of cases, might spread to other parts of the world.

The human immunodeficiency virus, or HIV, causes the most serious new pandemic of our time, which researchers discovered in 1984. By the end of 1997, an estimated thirty million people or more had been infected with HIV; nearly twelve million had died of AIDS because of their infection. By the year 2000, the number of people living with HIV/AIDS will be close to forty million. The most heavily affected continent is Africa, where HIV infects twenty-five percent of adults in some communities. Epidemics of AIDS are increasing sharply in several countries in Asia and Europe, and HIV infection is increasing in all parts of the world.

Unfortunately, new diseases are not the only causes of epidemics and pandemics today. Several older diseases have resurged, causing more cases and claiming more lives that they did years ago. Malaria has increased in Asia and the Americas, and the disease strikes 300 to 500 million people worldwide each year. In the Americas, the inci-

Significant Outbreaks of Smallpox in New World Peoples, 1507–1634.

When Europe began colonizing the New World, indigenous peoples were exposed to new diseases such as smallpox. Historically, the global movement of people has significantly altered disease patterns. Dates are from *Princes and Peasants: Smallpox in History* by Donald R. Hopkins.

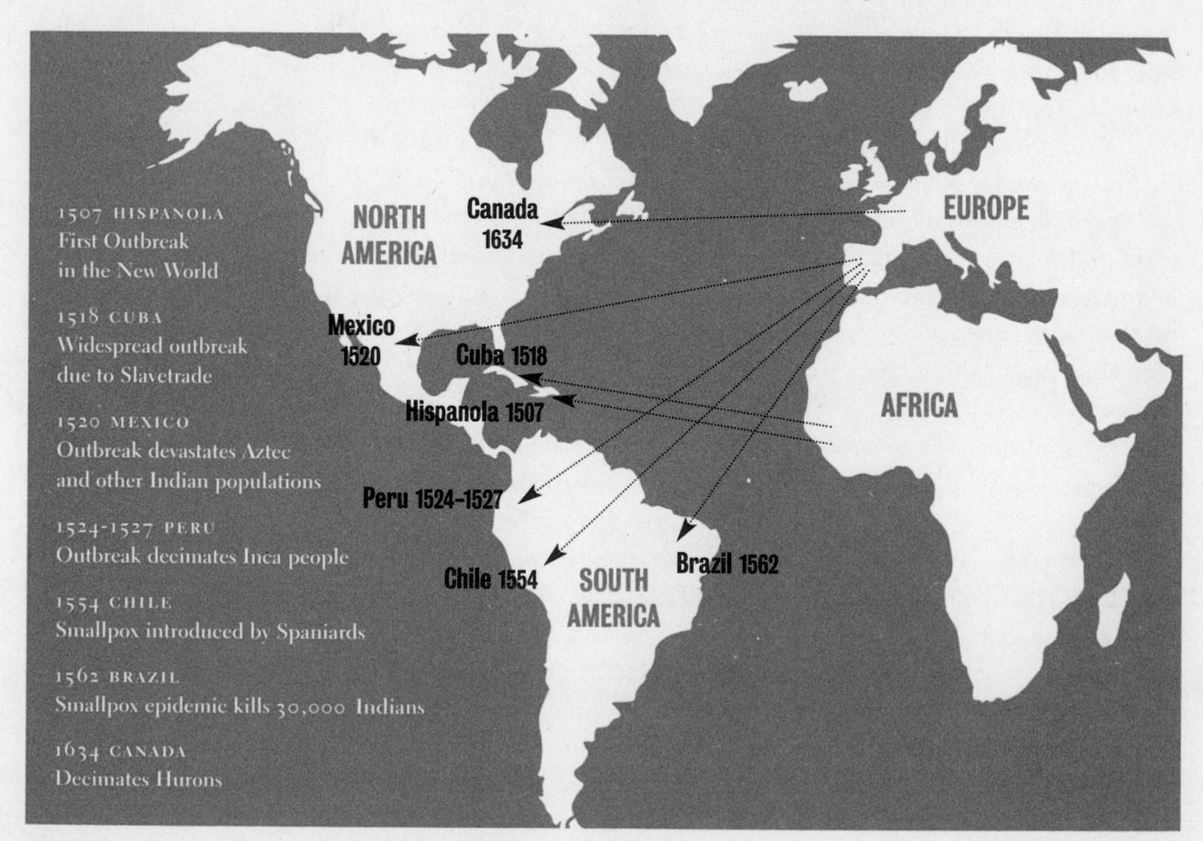

dence of dengue fever had declined to low levels through an intensive effort in the 1950s and 1960s to eradicate the mosquito that transmits the disease. Because of the inability to sustain eradication efforts, large epidemics have occurred since 1981; the tragic result has been hundreds of thousands of cases and hundreds of deaths from hemorrhagic complications of the disease.

We must always keep in mind that even common diseases can be deadly when they become epidemic. Periodically, influenza increases and causes deaths among older persons and those with underlying diseases. At the end of World War I, an influenza pandemic spread throughout the world, probably carried by military personnel returning to their home countries. The disease killed more than twenty million people, more than twice the number of military personnel and civilians killed during World War I. For that reason, there was considerable concern when a new type of flu, the bird flu, infected people in Hong Kong during 1997. Fortunately, the cases were few, and the disease did not spread beyond Hong Kong. However, the potential remains for influenza to spread rapidly and cause deaths in many people.

Epidemics of infectious diseases have affected humans since they began to congregate in communities and interact with other communities through war and commerce. Epidemics and pandemics became more serious as population and trade increased. The eradication of smallpox in 1977, one of the great epidemic diseases, shows that the prevention or elimination of many epidemics may be feasible. Under the leadership of the World Health Organization, the countries of the world are working to eradicate poliomyelitis (polio) early in the twenty-first century. Through the Pan American Health Organization, the countries of the Americas have already eradicated polio from the Western Hemisphere and are working to do the same with measles. These successes will lead the way to eradication of other infectious diseases. However, humans will probably continue to live with epidemics of infectious disease for decades to come.

What has been the impact of the development of antimicrobial drugs and vaccines in the treatment and prevention of infectious diseases?

How can an understanding of our long history with infectious disease inform the steps we take in the future?

Unsafe drinking water can contribute to outbreaks.

The Resurgence of Infectious Diseases

ANNE PLATT MCGINN

In May 1993, a physically fit twenty-year-old Navajo Indian—a cross-country and track star—began gasping for air while driving to his wife's funeral near Gallup, New Mexico. For several hours, the man suffered from an apparently severe case of the common flu. Then, abruptly, his condition worsened. Blood filled his lungs. He died in the emergency room, drowning in his own serum. Around the same time, three other healthy Navajos in the same region died of apparent cases of flu or pneumonia gone suddenly awry. Clearly, something horrific was on the loose—but what?

Intensive medical sleuthing revealed that the victims had been infected by the deadly hantavirus (named after the Hanta River in Korea where it was originally discovered); this virus severely damages the breathing tract and lungs. Investigators found that the carrier of this virus was the rural deer mouse, *Peromyscus maniculatus*, which is a native of most of North America, including the American Southwest. It was not immediately clear, however, why this normally shy animal had suddenly begun to appear and leave its droppings—in kitchens and playgrounds.

The disturbing hantavirus outbreak of 1993 is part of a larger pattern that involves a growing list of illnesses—and growing risks to hundreds of millions of people. At the end of a century in which infectious diseases once appeared to be under control, they are breaking out all over the world. Some pathogenic microbes, such as the *Escherichia coli* O157:H7 bacterium and the viruses that cause hepatitis C and Rift Valley fever, are new and unfamiliar. Others are old ones that we thought we had beaten, such as the microbes that cause tuberculosis, malaria, the plague, and measles.

AN EPIDEMIC OF EPIDEMICS

Although most infectious diseases are curable when treated quickly and properly, they nevertheless kill more than 17.3 million people each year, and their incidence is on the rise. Worldwide, tuberculosis kills three million people each year, while another

Anne Platt McGinn is a Research Associate with the Washington, D.C.-based World Watch Institute, a global environmental research group.

two million die of malaria, predominantly in tropical regions. For every person who dies, more than one hundred are infected. One third of the world's people—some 1.8 billion—now carry the tubercle bacillus, the bacterium that causes tuberculosis in susceptible people. Tropical diseases such as malaria, sleeping sickness, river blindness, and schistosomiasis infect more than 500 million people.

Despite natural disturbances, the environment in which the human species evolved and developed its basic defenses against disease remained stable for thousands of years. However, during the past 100 years, the Earth's physical and social environment has changed with unprecedented rapidity; this has allowed infections to spread much faster than anyone can develop the means of preventing and treating them. The depletion of forests, contamination of water, destabilization of climate, and explosion of urban population have all contributed to the weakening of public health protections. As a result, transmissions of infectious diseases through all media—air, water, insects, rats, and the human body itself—are on the rise.

Haunting these environmental changes is an ever-present biological pattern; rapid disruption always seems to favor opportunistic, short-lived species over stable, long-term ones. Among microbes, as among larger life forms, there are opportunistic varieties of bacteria or viruses that invade human blood or cells. Just as weeds exploit disturbance more quickly than slower-growing trees of a stable forest ecosystem, or insect pests exploit the reduced biodiversity of a crop, infectious agents can adapt fast enough to overwhelm societies whose "natural" environments are disturbed. An undisturbed ecosystem imposes a set of checks on the growth of microbes, but during severe disruptions the balance may be skewed in their favor. The more disruption there is in the human habitat, the bigger the biological risks are for the people.

BIOLOGICAL MIXING

The great increase in human tourism, migration, and trade has caused microbes that were once hidden in remote locations to have access to large human populations. Rapid settlement of the Amazon basin, for example, contributed to the spread of malaria-carrying mosquitos. The paving of the Kinshasa Highway across central Africa gave a fateful boost to the outbreak of AIDS. The sheer volume of traffic between different ecosystems has caused planet-wide biological mixing.

One reason the spread of microbes has been hard to stem is that they are invisible stowaways, carried unwittingly by their host; we have no way of knowing where the next outbreak will occur. However, any drastic environmental change increases the likelihood of such an outbreak.

After the Indira Gandhi Canal was built in Rajasthan, to irrigate desert-like areas of India, farmers switched from cultivating traditional crops of jowar and bajra to more commercially profitable wheat and cotton, which require large amounts of water. Many people came to the area in search of work. Then the monsoons came. Unfortunately, the main canal served as an ideal breeding site for mosquitos. Instead of high crop productivity and prosperity, the heavy rains brought the farmers tragedy and death: an epidemic of cerebral malaria, caused by a parasite carried by *Anopheles* mosquitos, quickly spread throughout extensive canal areas.

Along with the 100 to 200 known dangerous microbes, another 1,000 or so may be "out there," according to Paul Ewald, a biologist at Amherst College in Massachusetts. "It's a lottery" about whether or not a pathogen will be introduced into the population. Understanding the various links between environmental disruption, microbial outbreaks, and health may be nearly impossible, but recognizing some general patterns will allow for better prediction and disease prevention. Four factors—water contamination, climate change, human actions that magnify natural diseases, and increasing social disruption—have been associated with such "unpredictable" occurrences as the Hanta and Gandhi Canal incidents.

BAD WATER

Many outbreaks have been linked to the degradation of natural systems, particularly of water. Infectious illnesses are widespread in areas with overburdened sanitation facilities and unsafe drinking water. Water-borne diseases caused by pathogen-laden human and animal wastes—including malaria, cholera, typhoid fever, hepatitis A, salmonella, and dysentery—strike about 250 million new victims annually. Water-borne infections, such as cholera and other diarrheal diseases, account for eighty percent of all infectious diseases worldwide and ninety percent of all infectious diseases in developing countries. The worldwide population growth is increasing the demand on scarce water supplies in many areas; therefore we can expect the incidence of water-borne infections to keep rising, unless we make water quality protection a top priority in the planning and management of all water-using activities—from irrigation and hydropower to the disposal of sewage.

BAD WEATHER

Second, many outbreaks seem to stem from climate changes: warmer weather can expand the range of vectors. The Netherlands-based institute, Research for Man and the Environment (RIVM), estimated the effect of a global mean temperature increase of three degrees Celsius in the year 2100: it would increase the epidemic potential of mosquito populations in tropical regions twofold and in temperate regions more than tenfold. The RIVM model estimates an increase in malaria cases of several million—with more than one million additional fatalities—in the year 2100. Yet, that would be the toll of just one disease among many. The current heavy reliance on the combustion of fossil fuels is significantly increasing the atmospheric concentration of the greenhouse gas carbon dioxide. This may greatly contribute to global warming, with its risk to public and environmental health. To address these threats, governments need to begin tracking the broad connections between epidemics, energy, and transportation policies.

HUMAN AMPLIFICATION

A third factor is that infectious outbreaks appear to follow on the heels of human activity that magnifies the effects of natural disturbances—whether of floods, storms, or earthquakes. For example, some experts believe the outbreak of plague in Surat, India, in September 1994 stemmed from the flooding of the Tapti River that summer, and an earthquake a year earlier. The quake had left the landscape devastated and thousands of people homeless. Emergency aid and medical supplies were flown in for the survivors. However, the effort was so successful that excess food had to be stored in warehouses, where rodents crawled in and feasted. The rodents reproduced quickly, allowing the pneumonic plague bacterium—harbored in the fleas that infest the rodents' fur—to extend its range greatly. During the following summer, the monsoon flooded the Tapti River and covered the poorest districts of Surat with three meters of water. Again, people were forced to leave their homes. The rodents, too, were forced to seek shelter on drier land. The crowding together of rats and people increased people's exposure to the plague bacterium. Although India was medically prepared to deal with water-borne diseases such as gastroenteritis, cholera, and dengue fever, it had no plans for plague. The disease had not been seen in more than forty years. What could have been prevented or controlled at an early stage became a financial disaster as well as a social disaster, as international airline flights to and from India were canceled and trade temporarily halted.

SOCIAL DISRUPTION

Finally, the world's growing problem of social disruption has amplified the vulnerabilities caused by increasing environmental disruption. With about thirty civil wars now taking place, the systems needed for prevention and treatment of disease have been repeatedly shattered—often opening the way for infections to spread unchecked. In Russia, health conditions are worsening because of the combination of unstable political conditions, deteriorating infrastructure, and the transitional economy. Poor hygiene and diet, compounded by inadequate food supplies and high levels of pollution, have brought an onslaught of ecological and human health problems. In the early months of 1994, Russia had twenty-two percent more cases of tuberculosis than in the same months of 1993. Measles rose by 260 percent and mumps by ten percent during the same period. In 1976, diphtheria had all but disappeared from the former Soviet Union thanks to childhood immunizations. However, it surged back in the 1990s, rising from 1,200 cases in 1990 to 15,210 in 1993. What has befallen Russia is echoed, in varying degrees, in India, in Latin America, and in the Four Corners area of the American Southwest. The pattern occurs repeatedly: disrupted environments increase biological stresses on humans everywhere; mobility and population expansion increase their exposure to opportunistic microbes; and political or economic disruptions prevent the application of known preventions or cures.

WRITING THE PRESCRIPTION

Stopping the world's growing "epidemic of epidemics" may not be possible until we integrate considerations of human health into all major human activities—including the planning of irrigation and dam projects, road building and transportation systems, agricultural practices, and extractive industries such as mining and logging. In the future, along with keeping ecosystems intact and minimizing habitat alterations, communities should require planners to prepare for the unanticipated consequences of development. They would do well to provide ongoing health education for their populations, especially in areas that are particularly vulnerable to environmental disruption. Moreover, individual nations need to coordinate with the World Health Organization and with each other to establish a reliable global surveillance system to provide early warning, monitor incidence, and coordinate response. Adequate medical supplies and complete treatment therapies can then be targeted to at-risk populations.

When we begin routinely to take health impacts on our industries and societies into account, the outbreaks of disease that now shock us will not seem so puzzling. When the hantavirus broke out in Arizona, for example, it was not as mystifying to some Navajo medicine men as it was to the medical specialists. The medicine men's traditions had taught them to see the interconnectedness of all living things. They observed that before the outbreak, snow melt cascading down to the valley desert below, combined with a spring of heavy rains, had reminded some of their elders of the years 1918 and 1933, when there had been similarly wet weather. In each of those years, there had been a disease. In each of those years, piñon trees produced an abundance of pine nuts. Mice had descended on the extraordinary harvest and reproduced tenfold in one season. The rains had then forced the mice out of their flooded burrows to scurry about above ground, looking for food and shelter and increasing their exposure to humans. Disease was what happened when the balance of life was upset. "When there is disharmony in the world, death follows," said one medicine man.

Essay adapted from Anne Platt McGinn, "The Ecology of Infection," *World Watch*, July/August 1995.

What evidence illustrates that disruption of environmental systems can lead to a resurgence of a particular infectious disease?

What preventative measures have been implemented to override the outbreak of infectious disease caused by changes in climatic weather patterns or geological disturbances?

Urbanization often produces overcrowded housing conditions, as in this scene from Rio de Janeiro.

World
TB Day
March 24, 1998

TB Cases
New York City, 1978-1997

In New York City the number of cases reported in 1997 was 55% lower than the number reported in 1992. New York City contributed 31% to the national decrease in tuberculosis between 1992 and 1997.

The World TB Day 1998 poster shows the increase of the incidence of TB in New York City by year.

Public Health and Epidemics in New York City

MARGARET A. HAMBURG

Many people outside the public health community imagine that in the last years of the twentieth century, advances in biomedical practice and technology have made public health programs obsolete. However, underestimating the importance of public health programs is unwise and potentially fatal. Despite the dazzling array of advances in biomedical technology and treatment strategies, it is public health measures that have made the greatest contribution toward reducing disease, increasing longevity, and improving the quality of life and the health status of our citizens. Continued significant gains in health status will depend on public health insights and programs.

Throughout its history, the New York City Department of Health has been a national leader in public health, whether championing sanitation and hygiene, clean water, milk purity, vaccinations, or other public health measures required to promote health and safety. Today, important public health activities undertaken in New York City continue the tradition of protecting the public's health in an era of daunting challenges, both old and new. These core functions, many of which were pioneered in New York City, and are today shared by virtually every health department in the nation, include surveillance and control of communicable and other non-infectious diseases, protection from environmental hazards, health education and disease prevention programs, disease control activities directed at specific patients, and often providing other clinical services for the poor and others who would otherwise not receive adequate health care.

Clearly, these public health activities may not occur in a laboratory, a doctor's office, or a clinic or hospital, but they are nevertheless critical to the well being of our citizens. All these efforts reflect public health's role as "physician to the entire population." They are essential for attaining our national and local health objectives, and collectively represent an extremely cost-effective element of national and local health strategies.

The need for a robust public health system to

Margaret A. Hamburg is the Assistant Secretary for Planning and Evaluation at the U.S. Department of Health and Human Services. Prior to this position, Dr. Hamburg served for almost six years as Commissioner of Health for the City of New York.

prevent, detect, and respond to the threat of infectious disease remains as important today as when the forerunner of the New York City Department of Health was first established in 1866 in the face of a cholera epidemic. In fact, recent years have witnessed alarming increases in infectious diseases that once appeared to be under control or eliminated, as well as the emergence of new, previously unrecognized infectious diseases.

TUBERCULOSIS

Among the most serious of these resurgent diseases is tuberculosis (TB), which has occurred in New York since the city's founding as Fort Amsterdam in 1626 and has plagued human populations since recorded time.

Once the leading cause of death, the incidence of TB began to diminish, first with improvements in hygiene and then more significantly in the 1940s following the introduction of effective anti-TB medication. As the disease declined across the country, however, overconfidence, shortsighted public policy making, and failure to provide enough public funds for anti-TB efforts caused a reduction in the very TB control programs that had been so effective. By 1988, decreases in public health investments during the preceding two decades had devastated the Department of Health's, and therefore the city's, TB-control capabilities.

With the added weight of increases in poverty and homelessness, continued immigration from countries with a high prevalence of TB, and the companion epidemic of HIV/AIDS, incidence of TB in New York City tripled from 1979 to 1992. The highest TB rates were found among poor African Americans and Latinos, and especially among individuals infected with HIV, whose suppressed immune systems left them more vulnerable to TB disease.

Besides the dramatic rise in the number of TB cases, the other major concern was the alarming increase in the proportion of patients whose TB was resistant to treatment by more than one drug (multi-drug-resistant TB or MDRTB), including the occurrence of outbreaks and elevated rates of MDRTB in many hospitals and prisons.

Even under the best of circumstances, TB is difficult to cure. Effective TB treatment requires a regimen of multiple drugs taken over a period of many months. Poverty, substandard housing, homelessness, drug abuse problems, or other concurrent medical conditions further complicate TB control efforts by making it even more difficult to treat patients until cured. When TB patients are inadequately or incompletely treated, the *Mycobacterium tuberculosis* infecting them may not be completely eradicated. Worse, the surviving organisms are those that had the greatest resistance to a drug or drugs used in the original treatment. Drug-resistant TB requires even more complex and lengthy treatment strategies.

For all these reasons, by 1991 and 1992, New York City's TB control situation looked bleak. Cases of TB had nearly tripled in fifteen years. However, in 1993, following the implementation of an aggressive TB control program by the Department of Health, we turned an important corner: the number of new cases of TB declined by fifteen percent from 1992, and has continued to decline steadily every year thereafter. As of 1997, TB cases in New York City had declined by almost fifty percent.

The decrease in the number of new TB cases not only represents remarkable progress in turning back the TB epidemic, it proves the effectiveness of the comprehensive TB control measures provided to all city residents known to have TB by the Health Department and public and private health care providers throughout the city. In 1992, the city initiated a planning process called the TB Blueprint, formulated by the Department of Health acting with other relevant city agencies, and supplemented by a major addition of local and federal funding.

The Blueprint set out three major public health priorities for TB control: to increase the treatment completion rate for those people with active TB; to prevent the spread of TB in facilities where people are crowded together (e.g., hospitals, shelters, and prisons) and to prevent further cases of active TB through screening and treatment of those TB-infected individuals in whom TB is most likely to become active.

These three goals guided many improvements in TB control efforts. For example, the Directly Observed Therapy program expanded dramatically, enabling health care workers to travel the city to assure personally that patients took every dose of their anti-TB medication. Better patient treatment and follow-up resulted from increased education and awareness, and because of markedly upgraded and expanded clinical services at the New York City Department of Health as well as other health care providers in New York City. Similarly, infection control procedures in hospitals, as well as in other congregate facilities such as homeless shelters and correctional institutions, were improved, decreasing opportunities for the spread of disease in these settings. In addition, screening measures were begun to better identify those in need of TB treatment and to reduce the risk of ongoing TB transmission in these congregate settings. In particular, individuals entering city correctional institutions were screened for TB and a state-of-the-art communicable disease unit was built at the Rikers Island correctional facility so that prisoners with suspected or confirmed TB could be appropriately isolated from other inmates in this medical unit until TB disease was ruled out or until they were no longer infectious following the initiation of therapy.

The decrease in TB case rates in New York City represents a major achievement in the fight against this curable and preventable disease. However, this decline in no way signifies an endpoint in our battle against TB. Considerable progress still needs to be made before TB case rates return to where they were in 1979, before the current epidemic started, and even further to reach our goal of TB elimination.

HIV/AIDS

From an old enemy, let us turn to a comparatively new threat: the epidemic of HIV/AIDS. Currently, several critical realities of HIV in New York City affect our ability to respond to this epidemic; these realities shape, and often impede, the development and delivery of HIV/AIDS service in the city.

The huge scale of the AIDS epidemic, and its overlap with other epidemics of drug abuse, TB, and poverty, have serious and evolving implications for organizational and community resources at virtually every level. In its second decade, the face of the epidemic has changed dramatically. Once regarded as a disease affecting mainly gay males, the burden of HIV/AIDS has been increasingly and disproportionately felt in poor, minority communities; it is also increasingly a disease involving women, children, and families.

Today, drug abuse accounts for an increasing proportion of HIV transmission and has much to do with the changing face of HIV/AIDS. Through the sharing of HIV-contaminated syringes and needles, drug abusers can become infected and can transmit the disease to their sex partners and children. In the early years of the epidemic, women made up less than eleven percent of AIDS cases in New York

City. In recent years, approximately one quarter of all cases have been in women, either as a result of personal drug use, or through heterosexual transmission from an infected partner, often as the result of intravenous drug use. Too often these women are unaware of their risk.

HIV-infected women can transmit HIV to their fetuses, or to their newborns through breast-feeding. However, mother-to-infant transmission has dropped significantly in the past few years, reflecting the rapid implementation of prenatal HIV counseling and voluntary HIV testing guidelines and the increasing use of AZT therapy to reduce perinatal transmission.

Other factors making our fight against HIV/AIDS particularly difficult include: the great diversity of needs/capacities across the many subpopulations and communities affected by the overlapping epidemics; the burdens HIV/AIDS have imposed on the intricate, weakened health care and social services infrastructure of the city; and the administrative complexity and rigidity in funding, planning, and service delivery.

A range of strategies are especially important if we are to deal with this epidemic effectively. These include: (1) prevention and education, which represent our key weapons against the epidemic; (2) improved access to care and services, including ongoing research to develop new and improved treatment strategies; (3) addressing the broader social and economic context in which HIV occurs; and (4) perhaps most important, the development of more effective drug-abuse prevention and treatment programs and policies. It is now abundantly clear that we must take a more realistic and aggressive approach to the relentless threat of the drug-abuse problem, both as it relates to the spread of HIV and in its larger dimensions and implications.

LESSONS LEARNED: IMPORTANCE OF THE PUBLIC HEALTH SYSTEM

New York City's recent experiences with TB and HIV/AIDS have taught us that emerging or resurging infectious disease agents are neither relics of the past nor hypothetical threats. They are very real, imminent dangers in this era when advances in modern medicine, particularly the advent of antibiotics and vaccines, have lulled us into an illusory complacency. These diseases are sobering reminders of the need to be vigilant.

Over the next decade and well into the future, our nation will no doubt be faced with many new and previously unrecognized diseases—just as we have with Legionnaires' disease, Toxic Shock Syndrome, hantavirus pulmonary syndrome and, of course, HIV/AIDS. But in preparing for the unexpected, we should be mindful that the next significant threat will as likely come, not from a newly recognized organism, but rather from existing organisms that emerge to gain access to new host populations as a direct result of human activities, as well as the environmental changes we produce.

The barriers that protect us from assault by new or changing infectious organisms are fragile indeed. Perhaps our greatest weapons for defense can be found among those core functions of public health described earlier in this article. Improving our public health systems to better detect, investigate, and monitor emerging or resurging infectious diseases must be an extremely high priority. Sadly, despite the enormous contributions of public health and its critical importance for health and safety, our national commitment to core public health functions has faltered. The experience of New York City with infectious diseases offers several important implications for public health policy overall.

First, we must not tolerate crippling compromises in our public health system. The services and strategies that we require must be available and effective on both the local and national level, and must receive stable and adequate funding. The consequences of insufficient funding are too costly in loss of life and human suffering, as well as unnecessary and avoidable medical expenditures, lost productivity, and other economic costs.

Second, we must take a global perspective in the surveillance and control of infectious diseases. In this era of international travel and trade, this issue is not solely a matter of stemming the suffering, wasted lives, and preventable death caused by infectious diseases wherever they may occur, but it is essential to our efforts to prevent and control infectious disease within our nation. Our local abilities to meet the challenges posed by a global reservoir of emerging infectious diseases are really issues of international stability and national security.

Infectious diseases will increasingly raise a range of public health, medical, social, and public policy issues that cross international borders. These challenge the public and private sectors worldwide to work together to improve surveillance, disease control, research, and communications.

Third, we need more and better partnerships across local, state, and federal lines for research, training programs, and surveillance and control of emerging epidemics and infectious diseases. In the current political climate that favors reduced public spending and downsized government programs, public health programs are in jeopardy. Few local health departments have the resources to support adequate infectious disease surveillance programs. Improved linkages on the state and federal level are crucial, and we must expand and strengthen our national capacity for disease-specific expertise, national reference laboratories, basic and applied research, and national guidelines for treating and preventing infectious diseases.

Finally, I want to touch on the issues of hope and necessity. Our experience in New York City has repeatedly shown us the importance of social will and political courage in helping us to attain our infectious disease control objectives.

In sum, to recognize and control the emerging infectious disease threats that imperil us, we must continue to rely on the tactics of resourcefulness and persistence that have served us so well throughout the long history of public health in New York City. Effective responses to the staggering array of infectious disease risks that we will face into the foreseeable future will make important differences to all of us and will affect our society in ways that are impossible to calculate. With guarded optimism, I suggest that we will be able to continue that tradition of response. For the sake of all our citizens, we must do no less.

Adapted from a keynote address at the Museum of the City of New York in 1995 and a speech for the National Council on International Health in 1996.

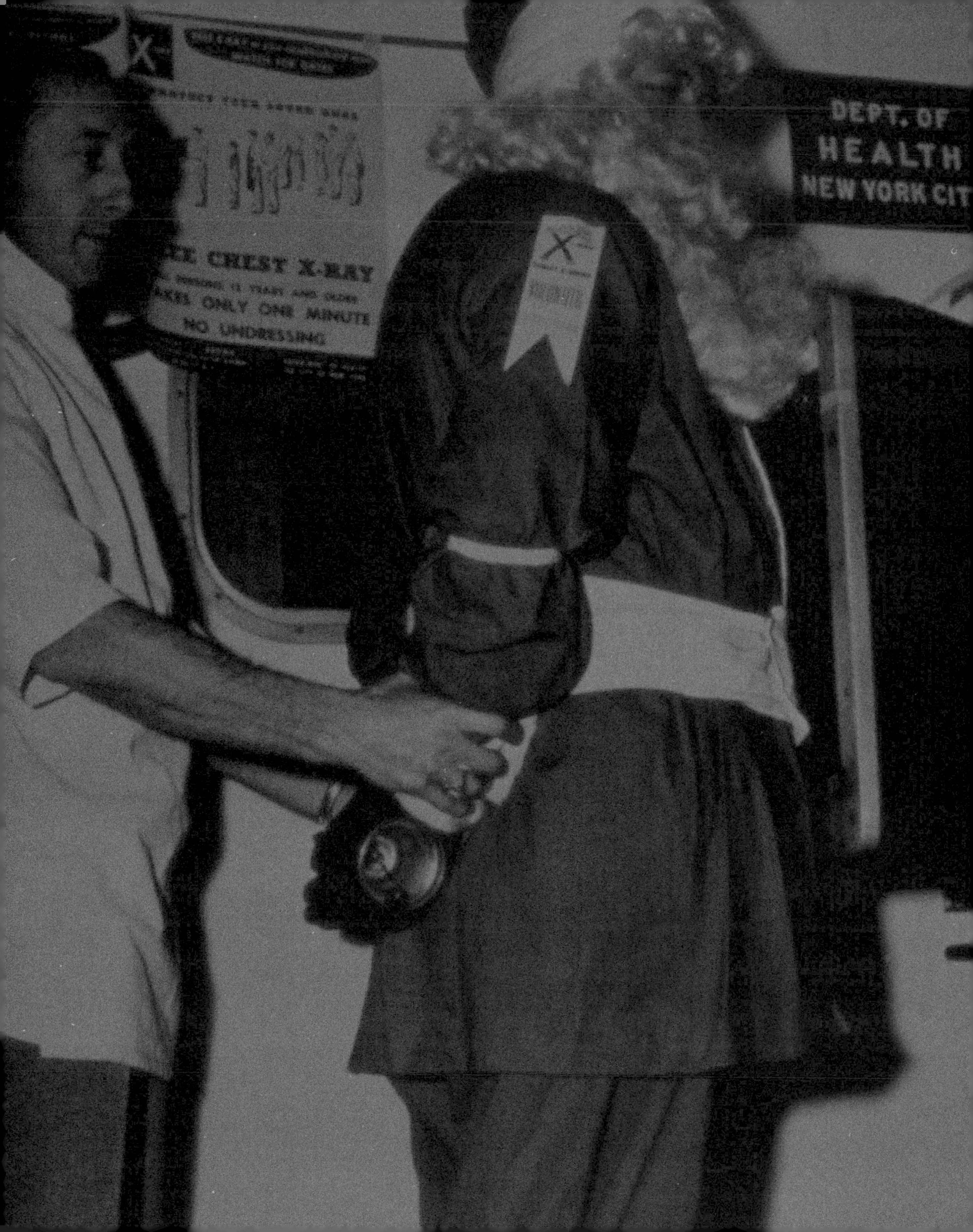
CHEST X-RAY
ONLY ONE MINUTE
NO UNDRESSING
DEPT. OF
HEALTH
NEW YORK CIT

How have the efforts of the New York City Department of Public Health demonstrated that preventative measures have been successful in curtailing infectious disease outbreaks?

left New York City Department of Health photo depicting Santa getting a TB X-ray.

How can the success in controlling TB be applied to the relatively new threat of HIV/AIDS?

In the effective treatment of infectious disease, what four strategies have been shown to provide the means by which diseases can be controlled?

Electron micrograph of HIV (*Retroviridae*).

The Impact of the African AIDS Epidemic

JOHN CALDWELL AND PAT CALDWELL

The contemporary AIDS epidemic ranks with other major plagues. In Europe, twenty million or more people died of the Black Death from 1347 to 1351, and globally perhaps twenty million died during the 1917–19 influenza epidemic. By the end of 1996, the world estimates for the AIDS epidemic were more than six million dead; a further twenty-three million were HIV positive, and therefore nearly certain to die of AIDS-related causes. Because of the open-endedness of the present AIDS epidemic, it will certainly kill more than other historic epidemics. The epidemics mentioned were over in three or four years. This relatively short duration was believed typical of epidemics.

In contrast, the AIDS epidemic is already a quarter of a century old, and the incidence of HIV infection is climbing both globally and in the Third World. Currently, when considering its impact on the populations in Africa, the AIDS epidemic can be likened to the plague. In its intensity, the epidemic is quite unlike anything experienced by national populations outside sub-Saharan Africa, although some sectors of other populations, such as homosexuals in the United States, may have comparable experiences.

THE AIDS BELT

The focus of the global AIDS epidemic is in sub-Saharan Africa, in a long belt stretching from the Central African Republic, Ethiopia, and southern Sudan through Uganda, Rwanda, Burundi, Kenya, and Tanzania to Malawi, Zambia, Zimbabwe, Botswana, South Africa, and Namibia. This belt, with about three percent of the world's population, has a majority (about fifty-five percent) of the world's HIV/AIDS cases.

In some of the cities of sub-Saharan Africa, the HIV/AIDS levels are unbelievably high: one quarter of the adult population. If the epidemic maintains its present level over the next few decades, most of the people in these cities will face death from AIDS-related causes. The epidemic in rural regions

John Caldwell is Emeritus Professor of Demography and Coordinator of the Health Transition Centre, Australian National University.

Pat Caldwell trained in anthropology and has been John Caldwell's constant collaborator since they began their anthropological demography work in the early 1960s.

may not be self-sustaining, but may depend on continued reinfection from urban areas.

The sub-Saharan African epidemic has the potential to change worldwide perceptions about AIDS because it is the only place where AIDS has almost exclusively afflicted heterosexuals. Homosexual and bisexual activity, and intravenous drug use are related to an estimated eighty-seven percent of HIV infection in the United States, eighty percent in Europe, and sixty-five percent in Latin America. In contrast, these risk factors account for no more than one percent of the sub-Saharan African epidemic. There, at least as many women as men are HIV positive, which results in massive infection of infants by their mothers. In terms of heterosexual transmission, there is infection among small numbers of girls from about twelve years of age, and among boys a few years older. The peak ages of death are in the twenties and thirties, thus removing the major earners in many families and orphaning many young children. Children often lose both parents because the parent first infected then infects the other.

MORTALITY

The obvious impact of the disease, and the reason it is so frightening, is that nearly everyone infected by HIV will die. However, strikingly unlike epidemics such as bubonic plague, influenza, or smallpox, not everyone has the same chance of infection. The sexually abstinent or those in a sexually monogamous relationship have no chance of being infected through sexual intercourse. Although extramarital sexual adventures occur, marriage provides some protection against infection. The predicted mortality from AIDS in the main AIDS belt is staggering; for example, in Kenya, present HIV levels mean that one third of those reaching adulthood is likely to die of AIDS. The rise in morbidity and mortality may all be ascribable to AIDS even if the cause of death is recorded otherwise.

Research on how people in the AIDS belt interpret and treat this disease reveals that about fifty percent of people with symptomatic AIDS blamed witchcraft. In contrast, only thirty percent of people suffering from other diseases attributed them to witchcraft. This helps explain why three quarters of all AIDS sufferers sought help from traditional healers who specialize in the occult. The percentage of those with other disorders who consulted these healers was much lower. Nevertheless, before death, half the people with AIDS had also visited hospitals, although only one eighth died there. Research suggests that most patients made these visits to get relief from distressing symptoms and pain, not with the expectation of being cured. The attempt to secure such alleviation impoverishes the families throughout the continent as they buy more and more medicines and pain killers.

FERTILITY

In the AIDS belt, women who are HIV infected are encouraged not to become pregnant and to practice family planning with the use of contraceptions, especially condoms. Recent research shows that HIV lowers fertility more than can be attributed to increased condom use; instead, it probably decreases fertility through biological mechanisms. If this is so, then population growth rates will decline more than previously predicted. The decreased fertility rate and increased mortality rate will probably result in very low rates of population growth in many eastern and southern African countries. Moreover, what researchers had viewed in some countries as the beginning of a voluntary transition toward a lower fertility rate, may instead be the effect of HIV on fertility. Unfortunately, attempts to predict demographic trends are greatly hampered by the inadequacy of the data collected on HIV levels in different segments of the population.

MARRIAGE

It is disappointing that little information has been collected on how AIDS has affected the institution of marriage in this region. Some evidence suggests that young women are putting off marriage because of fear of AIDS. Women of reproductive age, left widowed by husbands who died of AIDS, may be less likely to remarry because it is feared that they are HIV infected. The common practice of a widow being inherited by the brothers of her late husband had long been in decline, but now it has become even less common because of the AIDS epidemic. To compound this problem, temporary unions are becoming more common.

ORPHANS

Because of increased parental deaths, much attention has been given to orphanhood. Most adults die of AIDS when they are under forty years of age, and are likely to have young, dependent children. Caring for AIDS orphans can be a heartrending task in a region where one third of these orphans are likely to die of maternally transmitted AIDS during their first years of life. The high levels of orphanhood resulting from the epidemic will place great stress on the community.

In sub-Saharan African life, the extended family dominates the nuclear family in the sense that children are expected to make little distinction between their parents and their grandparents, their mothers and their aunts, or their fathers and their uncles. Relatives usually become the foster parents and, extraordinarily, the available evidence suggests that this fostering system will probably care for the very large number of orphans. Though exceptions exist, the extensive network of orphanages envisioned by some Western observers may be unnecessary.

THE SOCIAL EFFECT

Perhaps the most extraordinary aspect of Africa's AIDS epidemic is its limited social and political effect. This is a disease that in several countries will be the cause of death of half the population. For example, by the year 2000, AIDS will lower Zimbabwe's life expectancy by twenty years, to the level of half a century ago, and by the year 2010 by thirty-five years. The situation in the cities approximates that of European cities during the Black Death. Yet, surprisingly, life in East and Southern Africa is not traumatized. Governments are not threatened by accusations of mishandling the epidemic. Not one protest demonstration has occurred. Life goes on in a surprisingly normal way. There has not even been any very marked change in sexual behavior, and society is not dominated by government demands that there should be. There is no paranoia and little in the way of new religions or death cults. In some ways, it is very impressive.

However, to the outsider, it is also unnerving, and needs some explanation. The explanation is not that the epidemic mostly assails marginal groups. On the contrary, it is most intense in the cities, where AIDS attacks the upper classes at least as much as the poor and uneducated. It is also not that Africans are accustomed to disaster and have, until recently, suffered from similar mortality levels so that the experience is not particularly startling. Most countries in the main AIDS belt have not suffered war or civil strife, and young adults have not experienced such mortality rates in living memory.

Part of the explanation to this reaction is the

belief that death is not the end, which is a product of both traditional religion and fervent Christianity or Islam. A persistent image in African poetry and a belief held by most people is that death will come at its proper or destined time. Death may be preordained or it may result from unnatural forces, usually connected to someone bearing malice toward the one who dies. In any case, most people believe that little can be done to avoid death. They think that worrying about it or restraining sexual activity may undermine one's confidence and weaken one's resistance to ill-fortune and disease.

Moreover, politicians remain skeptical about foreigners' statistics and projections. They are also often ambivalent about whether sexual behavior can or should be changed. Much of the society, molded by the institution of polygyny, does not believe that only one woman can sexually satisfy a man. These sexual mores arise from the traditional religion's identification of female virtue with fertility rather than with virginity. In any event, the low level of demand by African governments for help has allowed the international community to respond halfheartedly to the epidemic.

THE FUTURE

The main AIDS belt is probably not typical of the whole of sub-Saharan Africa, and not a foreshadow of a situation likely to develop in the whole region. A solely heterosexual epidemic is not easy to sustain, and no such epidemic exists outside sub-Saharan Africa. The chance of transmission in one sexual act between two otherwise healthy persons except that one is HIV positive is low.

The sub-Saharan African epidemic appears to result from an unusual combination of circumstances. It depends on: (1) a considerable level of premarital and extra-marital sexual relations, often with parallel partners; (2) much of the non-marital male sexual activity being with prostitutes, partly because there is a widespread economic component in sexual relations and partly because of the substantial level of male migration that keeps them far from their wives; and (3) a poor health service that leaves many sexually transmitted diseases, which act as co-factors or catalysts for the transmission of HIV, untreated .

There is increasing evidence that another factor is important when added to the preceding three factors. This factor is that the main AIDS belt has almost two hundred million people of various ethnic groups whose males generally still follow the traditional custom of remaining uncircumcised. There is strong epidemiological evidence linking the incidence of HIV infection to ethnic groups not practicing male circumcision. In the early 1990s, this hypothesis predicted that southern Sudan, northern Mozambique, Botswana, and Namibia would witness a major rise in the prevalence of HIV infection. That has now happened, and it has occurred nowhere else. By tracking the relationship between HIV infection and circumcision, it seems likely that the main AIDS belt is now fully constituted.

Predicting the future course of the African and global epidemic is fraught with risk, but the following are a few hopefully informed guesses. The main AIDS belt will expand little or no further, and national HIV positive levels as high as ten percent will not develop elsewhere in Africa or the world. No epidemic comparable to the main AIDS belt epidemic will develop elsewhere in the world, even though elsewhere HIV epidemics are more often catalyzed by much more infectious homosexual transmission and/or intravenous drug use.

The most pessimistic conclusion is that progress in conquering the African epidemic may be decades away. The epidemic is already at least two

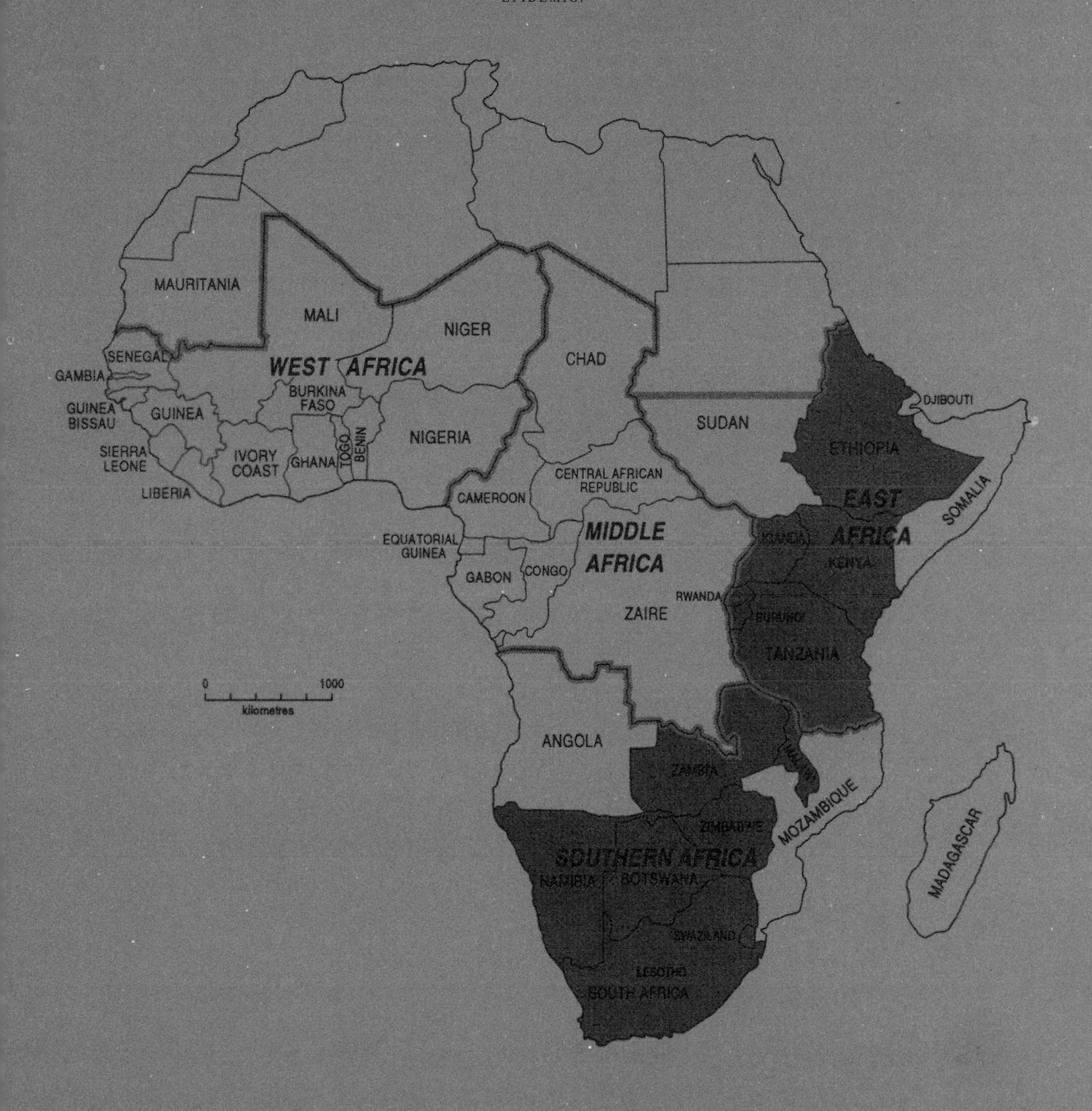

Dark areas indicate African countries with high incidence of HIV infection in 1997.

decades old, and HIV positive levels in most countries of the main AIDS belt are still rising. There is some evidence that HIV prevalence levels may at last be slowly declining in Uganda, but only among people less than twenty-five years of age.

As the epidemic stabilizes, its full effect will become clearer. For much of the main AIDS belt, the majority of deaths will be caused by the disease. One set of projections suggests that by the year 2010, Zimbabwe will have a life expectancy of thirty-five to thirty-six years instead of the current seventy years. This means that the worst is yet to come. So far, AIDS has killed more than four million people in the AIDS belt. Within a decade, four million may die every two years, at the rate of ten million per decade. These deaths will be a grim product of the continued rise in HIV levels during the 1990s.

When will the epidemic begin to pass? Less change in sexual behavior has occurred than most of us anticipated a decade ago, and there have been no massive national efforts to achieve such change. Nevertheless, one might conclude that there are gradual behavioral changes, which will probably eventually contain the epidemic. The most decisive will be the growth of cautious behavior in sexual relations, but this will probably be supplemented by an increasing use of condoms, especially outside marriage and other unions. Behavior is changing partly because of the impact of informational programs and news reports in the media, although the latter are on a smaller scale than might be anticipated. The main cause, however, is undoubtedly prolonged association with AIDS deaths through the loss of friends and relatives, and attendance at funerals. There will probably be some lasting effect on sexual mores, but the epidemic is unlikely to cause widespread monogamous, sexually exclusive marriage.

Even if the epidemic is contained solely by behavioral change, it will probably last for several more decades and kill in the main AIDS belt alone fifty million or more people. There may be a different scenario if there is a major medical breakthrough, such as an affordable and effective vaccine. This is by no means certain. Meanwhile, more limited interventions may help to limit the number of deaths. Some attack is being made on sexually transmitted diseases that act as co-factors, although here again the scale of the effort is more modest than might have been anticipated. Programs have begun in Africa on the use of drugs to reduce transmission from mother to fetus. There is much more doubt about whether the drug "cocktails," now becoming available in the West to prolong the latency period before HIV infection becomes full-blown AIDS, will be cheap enough or able to fit in with daily African life, to have any sizeable impact.

Though we have some knowledge of the demographic, social, and political effects of the AIDS epidemic, we must acknowledge that it will be many years before we can accurately assess the long-term impact on life in sub-Saharan Africa.

What circumstances in sub-Saharan Africa are responsible for the rapid spread of HIV infection and are not evident elsewhere in the world?

How does the AIDS epidemic change daily life in sub-Saharan Africa?

What steps could be taken to alter the present course of this epidemic?

Profile

Anthony Fauci: Leading the Offensive in the War on Infectious Disease

In the war against AIDS, Anthony S. Fauci has been the right man in the right place at the right time. As Director of the National Institute of Allergy and Infectious Diseases (NIAID) since 1984, when AIDS first hit the radar screen, Dr. Fauci is the commander-in-chief of the federal offensive against this deadly disease. NIAID is part of the National Institutes of Health (NIH). It engages in and funds research on AIDS and other infectious diseases, including malaria and tuberculosis, as well as asthma and allergic disorders.

A pioneer in the study of the physiological mechanisms controlling the human immune response to infection, Dr. Fauci has made several major contributions toward our understanding of HIV (human immunodeficiency virus). He discovered the means by which some of the body's own cytokines (chemical-messenger proteins), which are normally involved in inflammatory and immunological responses, allow HIV to replicate more efficiently. He discovered that HIV continuously replicates in lymphoid tissue at all stages of disease. Further, he demonstrated that a reservoir of virus remains in a latent form in resting cells even in individuals whose plasma virus has been suppressed by antiviral therapy. He has also developed therapeutic strategies to diminish the size of this latent reservoir.

Dr. Fauci has been at NIAID for most of his professional life. He started there in 1968, two years after receiving his M.D. from Cornell University Medical College, serving first as a clinical associate in the Laboratory of Clinical Investigation. He rose to the level of senior investigator and finally head of the Clinical Physiology Section before becoming Deputy Clinical Director of NIAID in 1977 and assuming the directorship seven years later. He also served as Director of the NIH Office of AIDS Research from 1988 to 1994. In addition to his administrative duties, he continues to engage in and direct research as Chief of NIAID's Laboratory of Immunoregulation.

Dr. Fauci is both a skilled advocate, frequently testifying before Congress in an effort to gain funding for research, and a hands-on scientist. In addition to his work in the field of HIV, he has made important contributions in the area of rheumatology, the study of autoimmune diseases.

From his command post at NIAID, Dr. Fauci keeps an eagle eye on emerging and reemerging infectious diseases, and he is concerned about our ability to combat them, especially in light of the development of drug-resistant microbial strains. He believes that surveillance and research are the two most important weapons we have. Infectious diseases must be tracked from their points of origin and epidemics closely observed. Basic research to increase our understanding of the microbes, especially the ways they change in response to drugs and other preventive and treatment measures, is no less important.

Dr. Anthony S. Fauci, Director of the National Institute of Allergy and Infectious Disease (NIAID).

"Despite our best efforts to prevent and control infectious diseases and immune disorders, changes in microbes and our environment will surely present new challenges," says Fauci. "The speed with which new and improved vaccines, diagnostic tests, drugs, and other measures can be developed will depend on our understanding of the human immune system and the ever-growing array of pathogenic invaders that threaten human health."

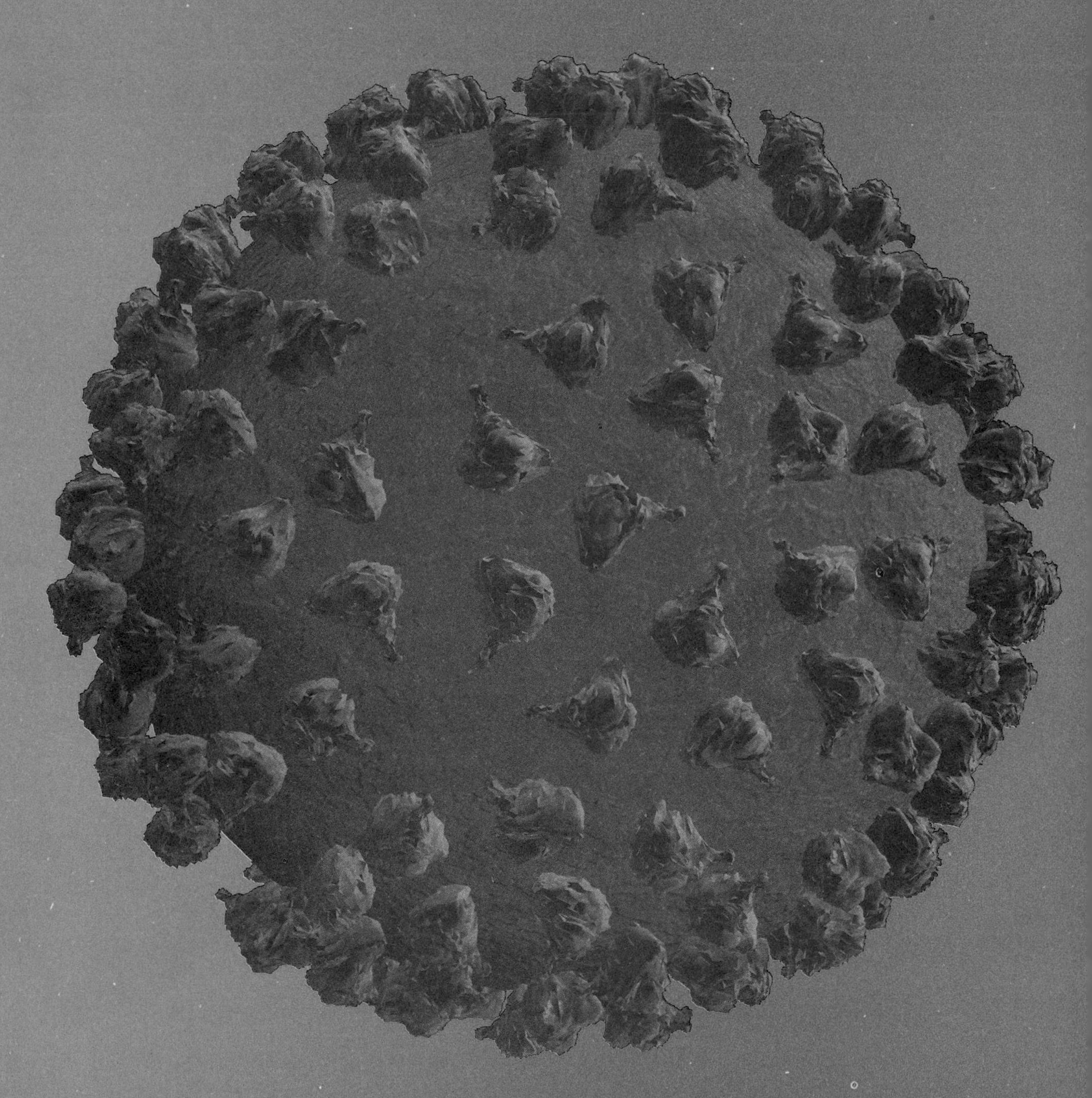

Computer-generated image of HIV *(Retroviridae)*.

Section Six: Action

City bus advertisement promotes condom use for AIDS prevention.

Section Six: Action

Introduction

ROB DESALLE

Infectious disease is a fact of life. Humans and microbes coexist, and there will always be conditions that lead to diseases of one type or another. Throughout history, people have developed many different ways to treat disease, from herbal remedies to the invention of vaccines. A greater understanding of the interaction between humans and microbes can help us devise more effective strategies for controlling, preventing, treating, and sometimes even eradicating diseases. Therefore, action represents the final concentric ring in our exploration of infectious disease.

Regardless of the type of disease or the cultural context, we are all united in our desire to reduce the devastating impact of disease. In the face of infectious disease, many people feel helpless. They want to know what they can do to protect themselves and their families from becoming ill. Those who are ill often feel isolated and alone.

In the last century, there have been great advances in the fight against infectious disease, and there are many actions that we can take to minimize the incidence of disease and the pain associated with it. René Dubos's well-known statement, "Think globally, act locally," provides a framework for structuring our actions. Dubos believed that the smallest changes in ecology, biodiversity, or human behavior can have far-reaching repercussions around the globe. A disease outbreak in Africa or Southeast Asia should not be ignored by people in the United States. All of the people on this planet share an ecosystem and should work together to find global solutions to threats or outbreaks of infectious disease. At the same time, local action is an essential component in solving global problems. Individuals and communities must work to change perceptions and policies that prevent education about disease and access to treatment.

Action can be achieved on several different levels—individual, community, national, and global. Any of these levels of action alone will not reduce the risk of disease, but if progress is made in each of these areas, we have a better chance of minimizing the damaging effects of disease-causing microbes.

On a personal level, there are direct actions that we can take to reduce our personal risk of exposure to microbes. For example, we can wash our hands, pour water through a sari, take medications properly (especially antibiotics), get vaccinations, and engage in only "safe sex." There are also indirect actions that a person can take to respond to infectious disease. We can volunteer in organizations that provide services to those who are ill. Through political and economic action, we can vote to enact laws and pass policies. We can purchase products and support companies that are making an effort to reduce the risk of exposure to disease-causing microbes.

On a community level, we can increase education and public awareness of an infectious disease. Involvement in charitable work is another important way we can take action, such as creating or supporting foundations that fund research or care for a specific disease. We can also work with service organizations that help those who are ill, such as hospices, needle exchange programs, condom distribution programs, support groups. In addition, we can support efforts like those of the World Health Organization's regional groups that monitor the health of residents in particular geographic regions and respond to their needs.

On a national level, we can work with governments to enact laws that provide for the safety of its citizens, such as food treatment guidelines, prescription drug practices, water treatment, and vaccination programs. A nation can provide social support to its citizens, such as access to medical care, medications, and immunizations. Nations can provide economic support, either for direct health care, prevention programs such as clean water supplies, or feeding programs to prevent malnutrition. Nations can also maintain surveillance programs of potential disease threats or existing diseases that may appear to be in decline.

On a global level, we can support organizations such as the World Health Organization that implement vaccination programs and other important health services. These efforts can help control disease and sometimes lead to eradication programs such as happened with smallpox and polio. Nations can put aside political and regional conflicts to work together to facilitate world health by sharing research and resources.

To explore action, I pose the following questions:

How does human behavior cause disease?

Medical and science writer LAURIE GARRETT describes how human behaviors can amplify infectious disease and have wide-ranging repercussions.

Can individuals really affect the course of an epidemic?

ALVIN NOVICK, Professor of Evolutionary and Ecological Biology, Yale University, addresses this complex question by looking at how people acted and reacted to the AIDS epidemic in the United States.

If human behavior can cause and amplify disease, can changes in behavior also minimize its risks?

LOUIS SULLIVAN, President, Morehouse School of Medicine, shows how an effective health initiatve like *Healthy People* 2000 can reduce the incidence of infectious disease.

This final selection of essays shows how human action and awareness at the global, national, and personal level can curb the spread of infection.

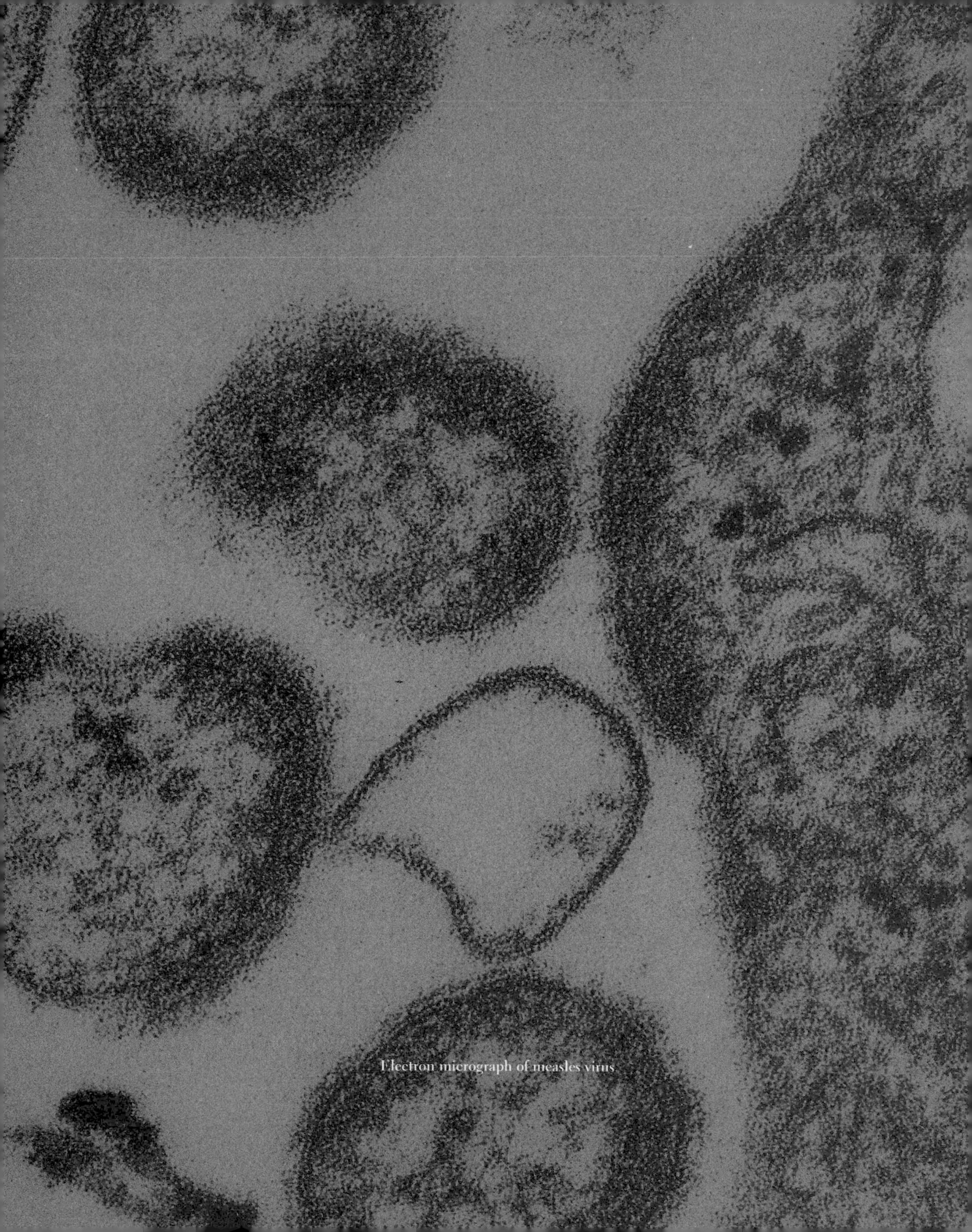

Electron micrograph of measles virus

Amplification

LAURIE GARRETT

If there is any wonderful news in the emerging diseases story, it is this: nearly all outbreaks and epidemics are the fault of our own species—of human beings—not of the microbes.

Such events as the great 1918–19 influenza pandemic, in which a particularly virulent and contagious strain of flu spread around the world five times in eighteen months, claiming some twenty-two million lives, are extremely rare. Though that tragic microbial event involved an unusually potent strain of the virus, human factors clearly played a role even in that flu pandemic.

It was World War I, and millions of troops were hunkered down in trenches stretching the length and breadth of Europe. Though the lethal flu strain is believed to have first surfaced among American soldiers in Kansas, the explosive spread of the virus followed transport troops to Europe, and their commingling with entrenched soldiers who came from as far away as Australia and New Zealand. Subsequent mass movements of WWI soldiers aboard ships and trains most certainly played the key role in expanding what was originally a modest Kansas outbreak into the worst global pandemic witnessed since the 1346 Black Death that claimed between a quarter and half of all lives it encountered from Mongolia to Tunisia, London to Mecca.[1]

When faced with the horror of microbial epidemics, it is tempting to throw up one's hands in fear and resign one's self to fate, as one might do for such natural disasters as earthquakes and tidal waves. So it is with relief that we realize that microbes generally spread by exploiting human behaviors—behaviors that may be changed or avoided, thus reducing or eliminating the opportunities for transmission of bacteria, viruses, fungi, and parasites.[2]

The starkest example of such human amplification of microbial emergence occurs with unfortunate regularity in South Asia and, more recently, Latin America, East Africa, and the former Soviet Union: cholera. The dangerous disease is caused by a simple bacterium, *Vibrio cholerae*, which lives in water supplies, human intestines, some types of shellfish, and human waste. Whenever sewer systems and drinking water supplies get cross-contaminated with bacteria, conditions are ripe for cholera epidemics. Such is the case annually in parts of

Laurie Garrett is a Pulitzer Prize-winning medical writer for *Newsday* and author of *The Coming Plague: Newly Emerging Diseases in a World Out of Balance*.

India and Bangladesh, where monsoon floods overwhelm inadequate local water and sewer systems, leaving millions of people exposed to contaminated floodwaters. More recently, cholera has broken out in Russia, Peru and most of Latin America, Tanzania, and Mozambique, because of a lack of chlorine or proper filtration. In such places, cholera literally poured from the water taps.

Certainly, issues of finances and international development complicate matters, but the cholera example illustrates a key point: epidemics are not inevitable and may be eliminated if *Homo sapiens* stop aiding and abetting their microbial foes.[3]

Perhaps the biggest vehicle of such dangerous humanly created microbial amplification is the needle. Whether used for injecting life-saving medicine into a patient's bloodstream or for mainlining heroin, needles and the syringes on which they are mounted constitute the most efficient method imaginable of transmitting microbes from one person to another.[4]

In 1933, a thirty-seven year-old man traveled from Iowa to Mexico, where he used a hypodermic syringe to inject narcotics—a syringe that he shared with other drug users. Upon his return to Iowa, the man came down with malaria—a parasite he had unintentionally injected along with the drugs.[5]

A year later an outbreak of malaria in New York's Rikers Island prison claimed forty-one men—again, the result of infections involving contaminated syringes.[6]

It is was widely known in medical circles during the 1930s that syringes could be vehicles for the spread of malaria, and the 1940s ushered in evidence of far more dangerous disease amplifications. In 1945, well before the virus for hepatitis B was discovered, it was known that outbreaks of the disease were related to reused medicinal syringes.[7] And by 1950, needle-spread outbreaks of several diseases were identified, including tuberculosis, syphilis, bacterial meningitis, yellow fever, and streptococcus pneumonia. Further, it was well documented that patients who required frequent injections for such things as diabetes or hemophilia were exponentially more likely to suffer from hepatitis and other blood-borne diseases.[8]

Yet syringes continued to be used carelessly, particularly by cash-strapped medical systems. As late as 1975, British physicians were still debating whether or not it was necessary to sterilize anesthesia needles between patients. And the Soviet Academy of Medical Sciences that year declared that it was safe to reuse syringes among many different patients so long as the needle tips were swabbed with alcohol.

In the northern Zairian town of Yambuku, a group of Belgian missionary nuns followed such advice, using just five syringes repeatedly on some 300 patients daily. When an isolated case of a bizarre illness showed up in that hospital in 1976, the entire population was swiftly in the grips of the world's first Ebola virus epidemic. Would that single first case have sparked an epidemic in the absence of such a hospital amplification system? No.[9]

Despite such clear evidence that misused syringes could expand outbreaks, even ignite epidemics of previously unseen microbes, medical systems all over the world continued to spread disease with dirty needles. In Singapore in 1979, for example, physicians spread hepatitis to 257 people.[10] Sicilian doctors sparked a pediatric hepatitis epidemic using contaminated syringes for 1982 school vaccine programs.[11] By the mid-1980s, critics were charging that injections and vaccinations were actually spreading polio in West Africa because of the reuse of syringes.[12]

And in 1986, researchers in Mama Yemo Hospital in Kinshasa, Zaire, offered proof that HIV was spread from patient to patient through reused syringes. Further, the bulk of HIV and AIDS in children at that time was due to the use of contaminated syringes for chloroquine injections, necessary to treat severe malaria cases.[13] Similarly, pediatric AIDS epidemics in Russia and Romania resulted from late 1980s reuse of contaminated hospital syringes.[14]

It soon became clear that HIV was also spreading among intravenous drug users—individuals who shared their equipment for injection of amphetamines, cocaine, heroin, and other drugs. The Thailand experience illustrated just how rapidly HIV could spread among IV drug users.[15] In 1987, just two percent of Bangkok's IV drug users were HIV positive. By January of 1991, more than ninety percent were infected. Similarly, HIV epidemics exploded in Scotland and India, propelled in just a matter of months by shared syringes among IV drug users.[16]

The breakup of the Soviet Union brought widespread bankruptcy to the region's health care systems during the 1990s. Needle reuse in hospitals became the norm, causing massive epidemics of hepatitis A, B, and C. Further, the high level of social alienation drove millions of youngsters to experiment with injection of narcotics, sparking epidemics of HIV and hepatitis.[17]

How much disease emergence could be eliminated if each and every injection was performed with a sterile syringe? No one knows for sure, but it certainly constitutes a highly significant percentage of all the world's disease amplification.

To stem the tide of HIV, some European countries distribute heroin injection kits among addicts so that multiple needle use does not occur.

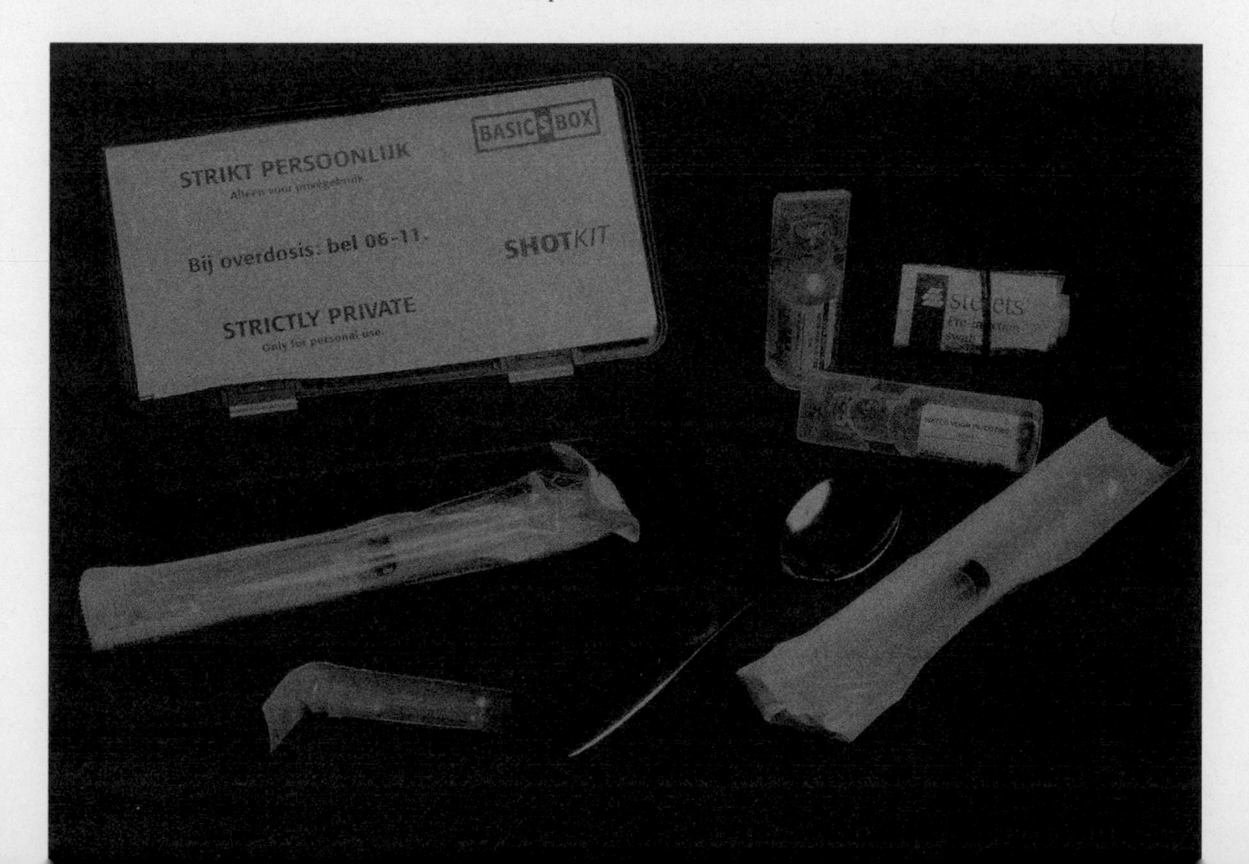

Other amplifiers worthy of consideration include contaminated drinking water, use of nonsterile surgical equipment, widespread unsafe sexual activity, global meat and food distribution, overcrowded jails and prisons, improper water drainage or, conversely, damming.

By focusing on human amplifiers of microbial emergence, solutions can be found. Facing the microbes need not elicit fear. Rather, common sense and political commitment can ensure that water is safe to drink, sewers remove and process all water, all syringe injections are sterile, condoms are used whenever individuals have promiscuous sex, the global food and meat supply is regulated, and crowded jails do not become vehicles for the spread of HIV and tuberculosis.

What evidence illustrates the role of humans in the amplification and spread of infectious diseases?

How can the transmission of diseases such as TB, hepatitis, and AIDS be curtailed?

1 S. S. Morse, *Emerging Viruses*, (New York: Oxford University Press, 1993).

2 M. E. Wilson, R. Levine, and A. Spielman, "Disease in Evolution," *Annals of the New York Academy of Sciences* 70 (1994).

3 L. Garrett, *The Coming Plague: Newly Emerging Diseases in a World Out of Balance* (New York: Farrar, Straus & Giroux, 1994).

4 C. E. Koop, P. Hartsock, and M. Samuels, "Public Health Tragedies Associated with Reused Syringes," First International Conferences on Self Destructing Syringes, New York, April 18, 1991.

5 O. C. Nickum, "Malaria Transmitted by Hypodermic Syringe," *Journal of the American Medical Association* 100 (1933): 140.

6 M. Helpern, "Malaria among Drug Addicts in New York City," *Public Health Reports* 49 (1934): 421–3.

7 Memo of Medical Officers, Ministry of Health.
"Role of Syringes in the Transmission of Jaundice," *Lancet* 2 (1945): 116–19.

8 L. P. Garrod, "The Nature of Meningitis Following Spinal Anaesthesia and Its Prevention," *British Medical Bulletin* 4 (1946): 106–8; R. C. Harris, L. Buxbaum, and E. Applebaum, "Secondary Bacillus Pyocyaneus Infection in Meningitis Following Intrathecal Penicillin Therapy," *Journal of Laboratory and Clinical Medicine* 3 (1946): 1113–20; P. V. Marcussen, "Syringe-transmitted Hepatitis in Venereologic Clinic," *Nordic Medicine* 40 (1948): 1760–63; P. M. Sherwood, "Outbreak of Syringe-transmitted Hepatitis with Jaundice in Hospitalized Diabetic Patients," *Annals of Internal Medicine* 33 (1950): 380–96; and M. Pestel, "Tuberculosis Transmitted by Syringe," *Press Medicine* 611 (1953): 551.

9 Garrett, op. cit.

10 K. T. Goh, "Hepatitis B surveillance in Singapore," *Ann. Acad. Medicine Singapore* 9 (1980): 136–41.

11 V. Intonazzo, G. LaRosa, A. Lanza, et al., "Epidemic of Viral Hepatitis in a Mountain Commune of Western Sicily," *Boll. Ist: Sieroter Milan* 62 (1983): 145–52.

12 H. V. Wyatt, "The Popularity of Injections in the Third World: Origin and Consequences for Poliomyelitis," *Social Science and Medicine* 19 (1984): 911–15.

13 J. M. Mann, H. Francis, F. Davachi, et al., "HIV Seroprevalance in Pediatric Patients 2–14 Years of Age," *Pediatrics* 78 (1986): 673–77; and J. M. Mann, H. Francis, T. C. Quinn, et al., "HIV Seroprevalence among Hospital Workers in Kinshasa, Zaire," *Journal of the American Medical Association* 256 (1986): 3099–102.

14 I. V. Patrescu and O. Dumitrescu, "The Epidemic of Human Immunodeficiency Virus Infection in Romanian Children," AIDS *Res. Human Retroviruses* (1993): 99–104; and L. Garrett, 1994, op. cit.

15 D. C. DesJarlais, S. R. Friedman, S. Vanichseni, et al., "International Epidemiology of HIV and AIDS among Injecting Drug Users," AIDS 6 1053–68.

16 Ibid.

17 L. Garrett, "Crumbled Empire, Shattered Health," *Newsday*, October 1997.

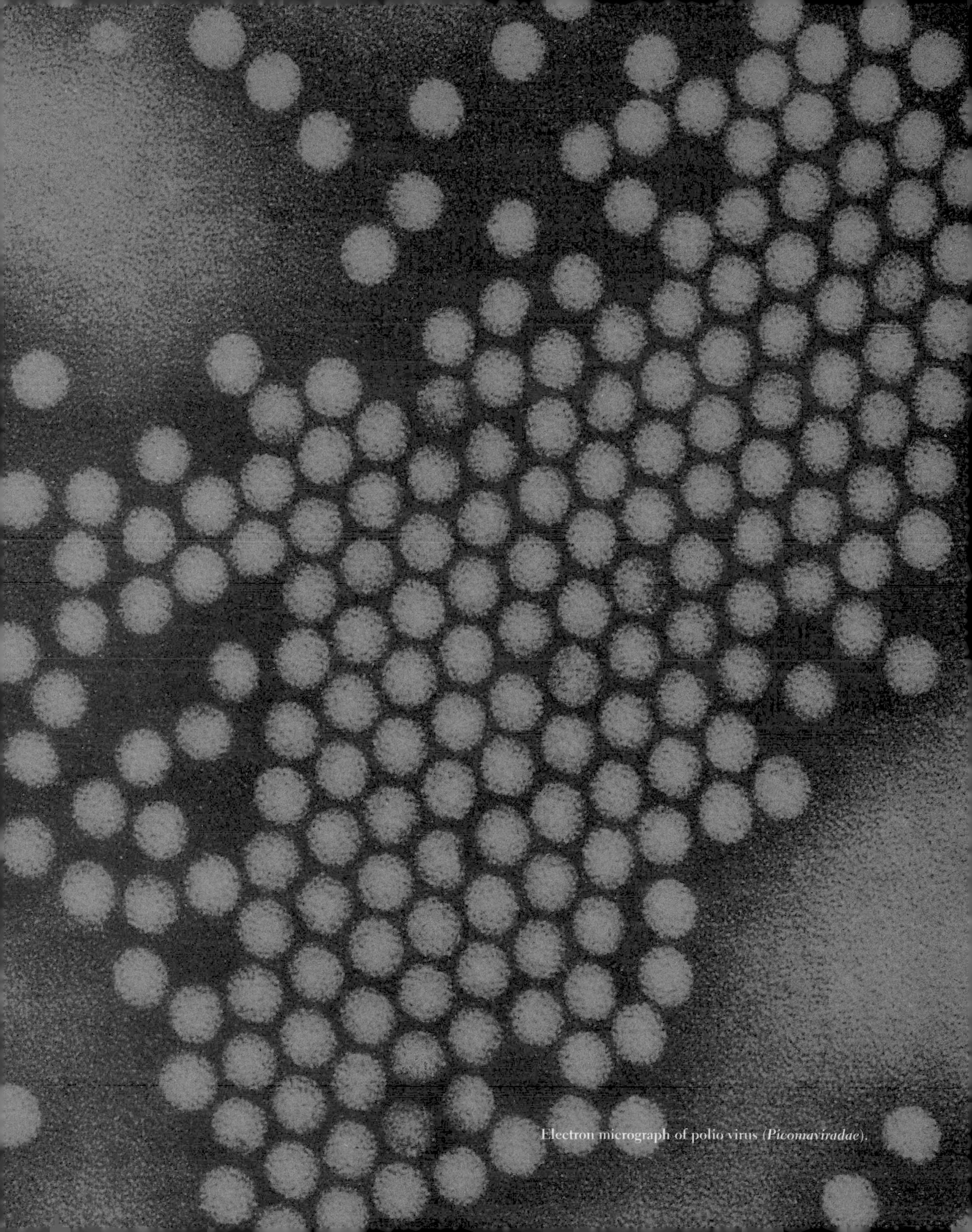

Electron micrograph of polio virus (*Picornaviradae*).

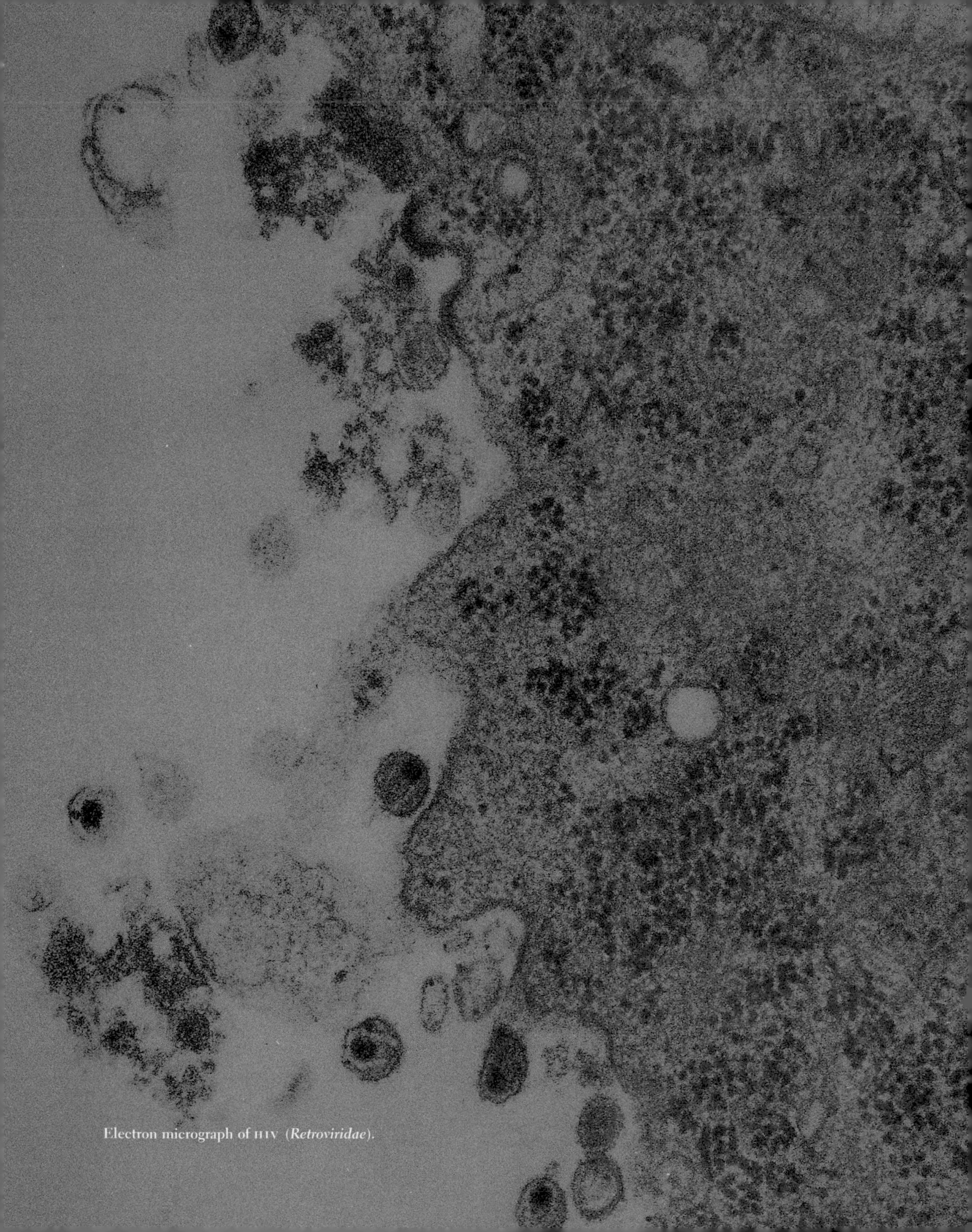

Electron micrograph of HIV (*Retroviridae*).

Can Individuals Really Affect the Course of an Epidemic?

ALVIN NOVICK

Many Americans presume that each of us, individually, lacks the power to influence the course of history. Yet, remarkably, the history of the HIV/AIDS epidemic, at every level, reveals that the fabric of the response was woven of multiple, identifiable individual efforts. These individuals defined the paths, invented the ways, and shaped our communities' and our nation's responses to AIDS. They are easily overlooked; many are only faintly recognized, but they put the bricks and mortar together of epidemiology and public health, community development, ethics, and public policy.

When what we now call AIDS was first reported in 1981, its incidence appeared to be associated with gay men. As our understanding of the new illness deepened, people who used illicit drugs by injection were also recognized as vulnerable to AIDS. However, most of the general public was still spared. Over ninety-five percent of persons with AIDS in the United States belonged and belong to social groups that were systematically ostracized and stigmatized long before the epidemic reached our shores. Many observers, including physicians and scientists, following the usual American disdain for gay men and "junkies," saw the victims of AIDS as guilty and unworthy. In consequence, there was a distancing from the epidemic, from the vulnerable people, and from the patients and their needs that persists in many ways even today.

Four events from 1985 typify the sluggishness of the American response to this crisis. At the First International AIDS Conference held in Atlanta under U.S. sponsorship, the Secretary of Health and Human Services stated that we had to stop the AIDS "scourge" before it reached the general public—but made no commitment to stopping the epidemic's spread among those who were at highest risk. In September 1985, Congress abruptly halted the first federal program to provide funding for HIV/AIDS prevention because this program provided information about "safer sex" through condom

Alvin Novick is Professor of Evolutionary and Ecological Biology and Public Health and Director of Law, Policy, and Ethics at the Center for Interdisciplinary Reasearch on AIDS at Yale University.

use. Similarly, the head of the AIDS program in a New England state received word that he would be summarily fired if he ever again publicly advocated condom use. Finally, the first conference on Ethics and AIDS, held at Cornell Medical School in September 1985, opened with a series of distinguished medical scientists telling the audience, in essence, "AIDS is overwhelming. Do not get involved!" Responding dramatically the following day, Helen Schietinger, who had created two successful AIDS programs in California, told the conference, "Yes, AIDS is overwhelming, but I simply provide a service." In these four cases, the official or "establishment" position was largely an effort to avoid facing up to the problem; however, many individuals were already "simply providing the necessary services" as best they could—often without adequate resources—and their programs inspired others.

By the late summer of 1981, Larry Kramer, a novelist and playwright and a gay man, had begun to raise money for the struggle against AIDS. He also began to organize what would become the Gay Men's Health Crisis—the first, the most influential, and now the largest community-based organization responding to AIDS in the world. Some of that history of the gay community's response to AIDS appears in Kramer's play, *The Normal Heart*. I do not wish to imply that Kramer, or any others that I will identify, did all the work themselves. They received aid, support, and encouragement from other individuals, many of whom faced risk because of their activism.

By late 1981, Dr. James Curran of the federal Centers for Disease Control had been appointed to initiate and head the national public health response to AIDS, a role he continued to play for a decade and a half. His sensitivity, professional skill, polished acumen, and decency in working with vulnerable groups shaped our nation's response to AIDS and frequently held the line against hostile reactions from politically conservative forces. The federal government's response was usually too little and too late but would have been far more meager had it not been for Curran.

At the local level, individual activists developed programs of information sharing, networking, buddy support systems, hot lines, meals on wheels, advocacy for extending health care coverage to AIDS victims, and AIDS prevention programs. These programs included advice on "safer sex" and, ultimately, the exchange of clean hypodermic needles for used ones to reduce the transmission of HIV via intravenous drug use. Initially most programs for preventing AIDS and for helping persons with AIDS grew out of the gay community. Later, other non-health professionals joined the fight against AIDS, usually in the face of opposition from vocal sectors of the public.

By the summer of 1983, representatives of perhaps a score of local AIDS service organizations, all with grassroots origins, met in Denver and created an overall national "umbrella group" to lobby, to educate, and to spread advice. As this group evolved, it split into two groups: the National AIDS Network and the AIDS Action Council. The former has disbanded; the latter is currently a powerful force in AIDS policy development. Each of the grassroots groups—and each of the umbrella groups as well—was largely dependent on individual advocates. These were volunteers who led, sacrificed, inspired—and often succeeded.

A new type of AIDS activism developed in early 1987—one often associated with groups calling themselves Act Up or something similar—initially in New York City and in San Francisco. The members of these groups were mainly gay and specialized in protests, civil disobedience, street theater, and other events that drew attention to the unaddressed needs of persons with AIDS and those most vulnerable to HIV infection. By the late 1980s and the early 1990s, New York's Act Up had evolved, first into the Treatment and Data Committee and then into the Treatment Action Group (TAG). Parallel evolution by different pathways led to Project Inform in San Francisco. Members of the latter two groups, led by a few inspired individuals, became citizen experts in virology, immunology, research design, clinical-trials design and analysis, and other scientific aspects of the response to AIDS. In consequence, their roles changed from mounting protests and serving as gadflies to participating in, and speaking at, conferences and to becoming members of policy-making committees and commissions. Mark Harrington and Greg Gonsalves from TAG, Martin Delaney of Project Inform, and their colleagues have strongly changed the course of HIV/AIDS drug development and of many aspects of conducting clinical trials and setting research priorities.

It was also individuals who led the way in developing needle exchange programs for intravenous drug addicts. In the Northeast, Jon Parker, then a student at Yale, began his advocacy for this class of people with AIDS or at high risk for HIV infection in 1983. By 1985, he had begun clandestine and informal needle exchange programs in several communities. In 1987, his efforts snowballed into a concerted effort to gain legal authority for a needle exchange program that could be rigorously evaluated to see whether it reduced the transmission of HIV among intravenous drug users. The New Haven Mayor's Task Force on AIDS undertook to create such a program, which three individuals organized: Sher Horosko, Elaine O'Keefe, and myself. Our individual successes in rallying support for a group of outcasts, in gaining the legal authority to set up a successful needle exchange program, and in scientifically evaluating the success of this program, significantly contributed to the subsequent establishment of similar programs in more than 100 U.S. cities.

Providing clinical care to AIDS patients is challenging and costly. In the early years of the epidemic, many of the nation's hospitals were reluctant to be seen as sites of skilled HIV care. For example, from the beginning of the epidemic Yale-New Haven Hospital had the status as Connecticut's source of expert AIDS care because it was the state's leading academic hospital and was associated with a great medical school; however, the hospital administration and the medical school faculty were reluctant to be "overwhelmed" by AIDS. The leading roles in the hospital's efforts against AIDS went to two nurses: Jeannee Parker (in 1983 and 1984) and Leetha Fraulino (from 1985 to 1990). Correctly assessing the size of the AIDS problem in New Haven, these two individuals fought fearlessly and tirelessly to develop what the hospital had initially envisioned as a very modest clinical AIDS program into one that was state of the art.

In cities with high incidences of AIDS, community-based responses to the disease originated in the gay male communities, often with the help of women health professionals. The response of other afflicted communities—black and Hispanic communites, heterosexual women, inner-city minority poor, prostitutes, the homeless, the mentally ill, and runaway or "throwaway" youth—came later but has been flowing at an increasing rate since 1985. Usually, the energy and vision of individuals has driven this response. For example, Yolanda Serrano in New York created ADAPT, which serves mainly blacks and Hispanics; Dominick Maldonado in New

Haven created Hispanos Unidos Contra el SIDA (Hispanics United Against AIDS) largely for Puerto Ricans; and Elsie Cofield, also in New Haven, created AIDS Interfaith, largely for African Americans. Comparable groups have sprung up across the nation, each typically catalyzed by one individual and aided, of course, by supportive colleagues.

The effect of individual vision, commitment, and historical consequence need not occur only in the direction of providing creative and effective service. Strong individuals can also block the development of good public health policy and thereby facilitate the epidemic, condemning thousands to death. For example, Gary Bauer, serving as a leading domestic policy adviser to President Reagan was apparently largely responsible for an official memorandum in 1987 stating that the Reagan administration would support only HIV prevention programs addressed to married couples, interestingly a population at almost no risk. No programs that addressed truly vulnerable people were to be federally funded. In a similar vein, North Carolina Senator Jesse Helms proposed an amendment, the so-called Helms amendment, to a late 1980s appropriations bill forbiding any federal funding for any purpose to any agency that directly or indirectly "promoted homosexuality." That amendment, which passed by an overwhelming vote of 96 to 2 (opposed only by Connecticut Senator Lowell Weicker and New York Senator Daniel Moynihan) remained in effect for six years and severely inhibited the development and carrying out of HIV prevention for gay men, since any reasonably effective description of gay "safer sex" could, of course, be seen as promoting homosexuality.

During the late 1980s, most supportive HIV legislation derived from the office of Rep. Henry A. Waxman of California, much of it was shepherded by Tim Westmoreland, his legislative aide. In the Senate, Senator Weicker was generally the leading proponent of effective AIDS legislation. President Bush created and appointed a Presidential Commission on AIDS chaired by June Osborn, an academic physician and public health professional. She chose as her executive director Maureen Byrnes (who had been Senator Weicker's leading aide for health care issues). This commission enlightened almost every aspect of the response to HIV/AIDS, and although the public and the president largely ignored its insightful reports, the guidelines then set are still leading the development of good policy.

Perhaps the most startlingly negative effect seemingly attributable to an individual is President Clinton's decision in 1998 not to allow any federal funding for needle exchange programs, although he acknowledged that scientific evidence supports the safety and effectiveness of these programs. We are, of course, used to profoundly important decisions being made by powerful individuals such as the president, but not when openly contrary to public health. There was barely a ripple of response to the president's decision except from the HIV/AIDS advocate community.

Contemporary history is not easily recorded, analyzed, or documented. Yet the now roughly twenty-year history of the response to the HIV/AIDS epidemic—surely the greatest current threat to health, life, and civil stability around the globe—has evidently been shaped, for better or worse, by individuals. Some of these have been powerful, but many have simply possessed vision, strength of will, and dedication to the public good.

How did individuals have an impact on the events that took place early in the battle to control the spread of AIDS?

What short- and long-term effects did the actions of various groups have on the AIDS epidemic?

Increasing AIDS awareness.

Profile

Mathilde Krim: Using Science to Fight Hatred

Mathilde Krim combines years of experience as a research scientist with deep personal convictions in her all-fronts battle against the deadly Human Immunodeficiency Virus (HIV) that causes AIDS. She has been a lifelong opponent of human injustice, in particular injustice rooted in ignorance that instills a fear of the different and the unknown. As cofounder and chairman of the board of the American Foundation for AIDS Research (AmFAR), she has dedicated herself to fighting the ignorance that has added the burden of cruel discrimination to that of people afflicted by life-threatening disease.

Growing up in war-torn Europe, Mathilde Krim saw what fear and hatred grounded in ignorance can do. She was born in 1926 in Como, in the northern Lake Region of Italy, and moved to Switzerland with her Swiss family when she was six. Despite living in a safe haven during the war years, rumors, at first, and then the reality of horrible persecutions inflicted on Jews and others, including homosexuals, profoundly affected Mathilde. After the war, she helped with the relocation of concentration camp survivors and other displaced persons. By then, she had already embarked on a scientific career as a biologist, studying at the University of Geneva from which she earned a Ph.D. degree in 1953.

Experimental sciences, she learned, teach not only facts but also honesty with oneself and humility. Progress—scientific progress in particular—requires open minds and the willingness, at all times, to accept new evidence and to adjust one's beliefs accordingly. "Because science teaches this, I believe that it is a profoundly civilizing factor," she says.

Through her experiences during and after the war, she developed a deep empathy with the Jewish people and studied their history, culture, and religion. She converted to Judaism, and she emigrated to Israel in 1953, just five years after the state was founded. There she was hired to work in a lab at the then-new Weizmann Institute of Science, where she studied cytogenetics and cancer-inducing viruses, research areas that led her to a life in America as well as, much later, to work on HIV.

At the Weizmann Institute, she met the American lawyer and motion picture executive Arthur Krim when he was touring the institute. They were married in 1958. She came to the United States and joined the research staff in the Division of Virus Research at Cornell Medical College in New York City. In 1962, she transferred to the Sloan Kettering Institute of Cancer Research, where she continued her work on viruses and, in the 1970s, set up and later headed a laboratory to study interferons, natural proteins produced by the body that fight viral infections and some cancers.

It was at Sloan Kettering, in 1981, that she encountered some of the earliest cases of AIDS. "Soon I realized that I was witness to the birth of a new and grave human disease," she says, and she was one of the first to understand that it would become a scourge with many and profound implications. "When self-appointed defenders of the 'public morality' started declaring that this disease was 'God's punishment for sin,' I threw myself to the defense of those who were not only so absurdly, but also so viciously accused."

As a respected medical researcher, she became a spokesperson to the media in an effort to educate the public about the infectious nature of HIV/AIDS, its modes of transmission, and its prevention. She formalized that effort in 1983, when she founded the AIDS Medical Foundation, that became, in 1985, the truly national AmFAR.

AmFAR's mission is to prevent death and disease associated with HIV/AIDS and to foster sound AIDS-related public policies. With support from the private sector, it has disbursed, to date, more than $150 million in grants to 1,745 research teams, focusing on vaccine research to prevent HIV transmission; treatments for HIV disease and its complications, as well as immune restoration therapies; and social research, policy analysis, and advocacy on legal and ethical issues related to HIV/AIDS, including the protection of the civil rights of people with HIV or perceived as at risk of acquiring HIV. AmFAR also advises federal and state legislatures, and it disseminates the information

Dr. Mathilde Krim., Co-founder and Chairman of the Board of the American Foundation for AIDS Research (AmFAR).

needed to continually improve the medical care of those suffering from HIV/AIDS.

Mathilde Krim's work for AmFAR embodies her lifelong dedication to both science and the battle for social justice. "I am proud that AmFAR has supported some of the fundamental work that has led to both the development of effective treatments and a greatly reduced rate of mother-to-infant HIV transmission. And I am particularly proud that AmFAR played a crucial role in the enactment of several milestones in federal AIDS-related legislation, notably, in 1990, the Ryan White CARE Act (with its AIDS Drug Assistance Program), and the Americans with Disabilities Act."

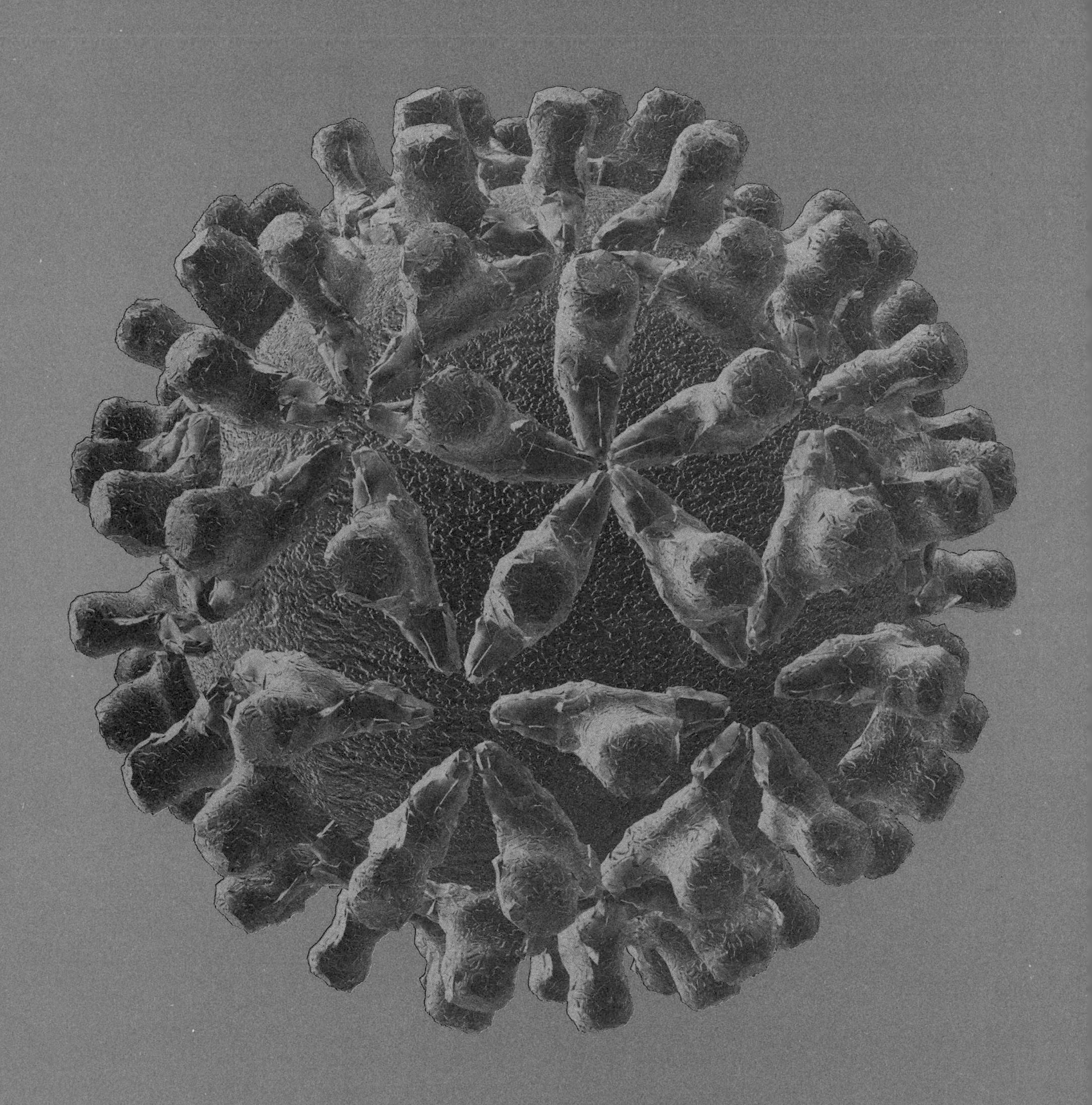

Healthy People 2000 and Infectious Diseases

LOUIS W. SULLIVAN

As the millennium rapidly approaches, we can look back on a century in which biomedical research greatly increased our ability to diagnose and treat a variety of disorders, including infectious diseases. Looking ahead, however, we can see a century in which an aging population, rising health care costs, and other economic trends will severely challenge our nation's ability to provide adequate health care for all its people. *Healthy People* is an evolving comprehensive agenda to help meet this challenge by substantially reducing the incidence of preventable diseases and injuries that significantly threaten the nation's health. Based on the best scientific knowledge, *Healthy People* provides a blueprint for public and private sector efforts to address these threats.

Healthy People originated in a 1979 surgeon general's report bearing the same title. The foreword to that report boldly proclaimed the agenda's revolutionary intent: to recast the nation's public health strategy to emphasize the promotion of health and the prevention of disease. Underlying this agenda was the belief that clearly stated health goals and strategies, with careful monitoring of our progress toward them, could lead to improvement in the health status of our citizens.

Healthy People 2000, published in 1990, is a comprehensive agenda with three main goals: lengthening the span of healthy life for Americans, reducing health disparities among different U.S. population groups, and providing all Americans access to preventive health services. These three goals comprise 319 objectives for the year 2000, organized into twenty-two priority areas; these are under the four general preventive-intervention areas of health promotion, health protection, preventive services, and surveillance and data systems. *Healthy People* 2000 also lists federal and national information sources concerning the twenty-two priority areas and suggests how to obtain additional information from state and local sources. Sharing lead responsibility for the priority areas are the Centers for Disease Control and

Louis W. Sullivan has been President of Morehouse Schools of Medicine in Atlanta, Georgia, since 1993.

Prevention, the Food and Drug Administration, the Health Resources and Services Administration, the National Institutes of Health, the Office of Population Affairs, the President's Council on Physical Fitness and Sports, and the Substance Abuse and Mental Health Administration.

Developing *Healthy People* 2000 involved private and public organizations, professionals, and nonprofessionals from all parts of the United States. Work on this agenda started in 1987 with the formation of the *Healthy People* 2000 Consortium; this alliance has grown to include 350 national membership organizations and 300 state health, mental health, substance abuse, and environmental agencies. Along with the Institute of Medicine of the National Academy of Sciences, the Consortium helped the U.S. Public Health Service convene eight regional hearings that received testimony from more than 750 organizations and individuals. Working groups of professionals used this testimony to develop the initial versions of the health objectives. Then, more than 10,000 people participated in the extensive public review that led to the refined and revised objectives in this agenda.

Healthy People provides a national basis for measuring performance by results. For the federal government, states, localities, and many private sector institutions and organizations, *Healthy People* is a strategic management tool. Improved health status, reductions in risk factors, and improved provision of certain preventive health services are the yardsticks that measure success. Progress reviews occur periodically for each of the twenty-two priority areas and for such population groups as women, adolescents, and racial/ethnic groups. Congress has designated *Healthy People* objectives as the criteria to measure the success of the Indian Health Service, the Maternal and Child Health Block grant, and the Preventive Health Services Block grant.

Taking *Healthy People* as a guide and model, to date, forty-seven states, Washington, D.C., and Guam have drawn up their own *Healthy People* agendas. Most states have included national objectives, but almost all have formulated them to meet their specific needs.

Development of a national health agenda for 2010 is well underway—based on *Healthy People* 2000—and refined by ideas and information provided by focus groups, public meetings, other forums, and countless individuals and organizations. *Healthy People* 2010, which is scheduled for publication in January 2000, will address such emerging issues as new medical technology, changes in the age profile and racial/ethnic composition of the population, advances in preventive medicine, and increased public awareness of, and demand for, preventive health services. Public involvement in the development of *Healthy People* 2010 will continue until September 1999. The first round of comments on *Healthy People* 2010 is now available at the Web site **http://web.health.gov/healthypeople**, which will continue to provide updated information on the agenda's development.

Recent advances in medical and environmental knowledge have taught us much about the contributing factors to many health threats. As individuals, and as a society, we can use this knowledge to decrease the incidence of preventable diseases and disabilities. Not only will this improve our own and our nation's health, it will free up medical and economic resources that can be directed against diseases and disorders that currently cannot be prevented. In providing an evolving agenda against preventable injuries and diseases, *Healthy People* will continue to play a crucial role in improving the nation's health and in enriching the lives of its citizens.

Priority Areas for HEALTHY PEOPLE 2000

The three priority areas of *Healthy People* 2000 most directly concerned with infectious disease are:

- HIV infection
- sexually transmitted diseases
- childhood immunizations and infectious diseases

These areas are within the general area of preventive services. Reviews of progress in these three areas—published between autumn 1996 and summer 1998—offer some encouragement but still paint a picture that is far from rosy.

HIV/AIDS

Positive news from the HIV/AIDS front includes:

- a continuing decline in the incidence of AIDS cases overall and among blacks and Hispanics
- a decline in the percentage of fifteen-year-old males and females who have had sexual intercourse
- an increase in condom use
- increased HIV testing at family planning clinics
- an increase in the number of HIV education programs offered at high schools, colleges, universities, and large and small businesses.

On the negative side:

- The incidence of AIDS in women is increasing (from 10.9 per 100,000 in 1994 to 11.2 per 100,000 in 1995).
- The percentage of IV drug users in drug treatment programs is declining (from 47.8 percent in 1994 to 34.1 percent in 1995).
- As of the summer of 1997, an estimated forty percent of those at risk of HIV infection had not been tested for infection.
- Tragically, AIDS—a completely preventable disease—is now the leading cause of death for people aged twenty-five to forty-four.

SEXUALLY TRANSMITTED DISEASES

Among industrialized countries, the United States has the highest rates of sexually transmitted diseases (STDS), which include HIV/AIDS, syphilis, gonorrhea, hepatitis B, chlamydia, pelvic inflammatory disease, genital herpes, genital warts, and infection by the human papillomavirus.

- During this decade, we have made great progress against gonorrhea and syphilis:

- The incidence of gonorrhea decreased by fifty-six percent between 1990 and 1997, from 300 per 100,000 to 123 per 100,000, the lowest rate ever reported in the U.S.
- The incidence of primary and secondary syphilis dropped to 3.2 per 100,000, the lowest recorded rate and eighty-four percent below the peak incidence in 1990.
- Among blacks, the decline in syphilis was particularly dramatic—from 118 per 100,000 in 1989 to 22 per 100,000 in 1997.
- The ratio of the incidence of primary and secondary syphilis in Hispanics to that in whites dropped from 6:1 in 1990 to 3:1 in 1997.

  In 1997, half of all reported cases of primary and secondary syphilis occurred in thirty-one (one percent) of America's counties, and seventy-five percent of all U.S. counties were syphilis-free; the latter statistic suggests that this disease could be eliminated nationwide.

The most commonly reported STD is chlamydia; this is a major cause of pelvic inflammatory disease, and a leading cause of infertility.

- In 1996, the states reported that fifteen to twenty-four-year-old females had incidences of chlamydia ranging from 2.5 percent to 10.9 percent. As high as these figures are, they probably are significantly below the actual numbers.
- The viral vector of genital herpes—herpes simplex virus-2 (HSV-2)—now infects one in five adults in America. The overall incidence of this viral infection has increased thirty percent since the late 1970s; among white teenagers, the increase has been 500 percent.
- According to current estimates, one half of sexually active adults—including thirty-four percent of college-age women—are infected with the human papilloma virus (HPV); HPV infection is the most important known risk factor for cervical cancer.

CHILDHOOD IMMUNIZATIONS AND INFECTIOUS DISEASES

We have made significant progress concerning Immunization and Infectious Diseases:

- The incidences of seven diseases preventable by vaccination—diphtheria, tetanus, polio, measles, rubella, mumps, and Haemophilus influenzae infection (for children younger than five)—reached record lows in 1995.
- Since 1993, all recorded cases of measles in the United States have involved people who were infected abroad; measles no longer is native to the United States.
- Between 1987 and 1995, the number of reported cases of hepatitis B dropped from 63.5 to 22.9 per 100,000.
- For hepatitis C, a drop from 18.3 to 3.7 per 100,000 occurred.

  The screening of blood supplies played a major role in reducing the incidence of hepatitis C, and needle-exchange programs may have played a role in the seventy-seven percent reduction of hepatitis B infection among IV drug users since the 1980s.

On the negative side:

- Thirty-three reported cases of hepatitis A per 100,000 is the same as in 1987—twice the target figure for the year 2000.
- Although there is a vaccine against pertussis (whooping cough), the incidence of this disease has increased throughout the 1990s; in 1995 there were 5,137 reported cases, more than five times the year-2000 target.

INFECTIOUS DISEASE DISPARITIES

Despite generally positive trends in the fight against infectious disease, disparities remain among groups with different income levels and among ethnic groups.

- Children from families with below-poverty incomes are less likely to receive recommended vaccinations than are children of families whose incomes place them at or above the poverty level: in 1996, sixty-nine percent of below-poverty children aged nineteen to thirty-five months received the recommended series of vaccinations; for other children, the figure is eighty percent.
- For elderly blacks and Hispanics, rates for immunization against influenza and pneumococcal pneumonia are only half that of the population as a whole.
- In most population groups, the incidence of tuberculosis (TB) has remained constant or declined—for an overall decline to 8.7 cases per 100,000 people in 1995 from the 1988 baseline of 9.1. However, among Asians and Pacific Islanders, the incidence increased from 36.3 (in 1988) to 45.9 (in 1995) per 100,000. There has also been a decrease—from 66.3 percent in 1987 to 65.3 percent in 1993—in the proportion of people diagnosed with TB who completed their course of preventive therapy. Such a failure to complete the course of preventive therapy is a major contributor to the development of drug-resistant TB strains.

How are the three priority areas of *Healthy People 2000* influenced by factors such as income, ethnicity, and education?

How can *Healthy People 2000* be used as a tool to assess, improve, and communicate our nation's health?

Profile

William Patrick: Waging War Against Infectious Disease... Literally

How often we hear about metaphorical warfare on infections and other diseases: the war against cancer, the battle against AIDS, the conquest of polio. But infectious disease plays a part in real warfare and a man named William Patrick has fought for both the offense and the defense.

Biological warfare is the use of bacteria, viruses, and other pathogens to attack enemy personnel or civilian populations, weakening their ability to attack or defend themselves from attack. It's germs instead of bullets and bombs.

The United States began developing biological weapons during World War II. In 1943, afraid that Germany would initiate germ warfare, the army established a top-secret laboratory at Fort Detrick, in Frederick, Maryland. In the end, none of the warring nations used biological weapons, but fears and research continued in the postwar years and intensified in the 1950s, during the Korean War.

William Patrick was born in 1926 in a small town in South Carolina. After serving in the army during World War II, he studied biology at the University of South Carolina, then did graduate work in microbiology at the University of Tennessee. In 1949, armed with a master's degree, he went to work for a pharmaceutical company, just as antibiotics were beginning to change the way medicine was practiced. He intended to devote his professional life to these new wonder drugs, but a phone call from one of his graduate school professors changed his direction.

The call was an invitation to join the germ warfare team at Fort Detrick. Patrick was intrigued, and by 1951 he found himself helping to build this country's biological arsenal. Soon, he was head of a development team assigned to design ways to grow pathogens, preserve them, and deploy them against an enemy. Ultimately, he became head of product development, in which capacity he was involved in tests of American bioweapons' capability. In carefully planned and closely monitored tests, airplanes disseminated selected germs over barges carrying hundreds of guinea pigs and rhesus monkeys, killing more than half of them.

In 1969, President Richard Nixon ordered an end to America's biological weapons development program amidst growing international sentiment against germ warfare as a moral means of waging war. The Fort Detrick facility was transformed into the U.S. Army Medical Research Institute of Infectious Diseases and its mission changed from an offensive to a defensive one. William Patrick stayed on at Fort Detrick for a while, but now works as an independent consultant, advising governments and law enforcement agencies on defending themselves against the threat of biological warfare.

The United States has never used biological weapons against humans and it is unclear whether any other nation has either. There are suspicions that germ warfare was used by Iraq against U.S. troops during the Gulf War, and there is little doubt that, despite treaties banning biological weapons research, other countries have stockpiles of pathogens for combat use.

It is thanks to William Patrick that American troops were vaccinated against anthrax before launching Operation Desert Storm in 1990 and through him that we know that Russia maintained stores of smallpox, anthrax, and plague pathogens. William Patrick's expertise has proven invaluable when it comes to countering the threat of biological weapons in the hands of terrorists and other rogue groups. In 1985, he helped trace an outbreak of salmonella poisoning in Oregon that sickened 750 salad bar customers to the cult led by Bhagwan Shree Rajneesh, which was waging a war against local community opposition to their activities.

What makes a pathogen a likely candidate for a biological weapon? According to William Patrick, it must be a reproducible organism that is not only infectious but remains so while being sprayed across a targeted area. And, he adds, it should be one for which treatment or a vaccine is available. If that sounds odd—wouldn't a weapon be more powerful if there were no way to combat it?—Patrick explains that it is extremely important to protect workers who may be at risk when creating and handling the weapons.

The pathogens he considers the greatest threats include anthrax, a bacterial infection carried by sheep and other ungulates (animals with hooves) that attacks the lungs in humans, leading to death if untreated; tularemia, a bacterium usually spread by ticks and infected birds and mammals, that causes severe flu-like illness (high fever, vomiting, headaches, extreme weakness) and sometimes pneumonia; brucellosis, a bacterial disease of pigs and cattle, that also resembles a severe case of flu; equine encephalitis, a mosquito-borne virus that causes fever and neurological symptoms (including tremors, confusion, convulsions, coma); and staphylococcus B, which can cause severe gastrointestinal distress and toxic shock.

How serious is the threat of biological terrorism? William Patrick says that whereas it would be difficult to contaminate a municipal water supply, for example, because purification systems designed to eliminate naturally occurring pathogens would also take care of any introduced maliciously, infectious agents and the tools used to spread them can easily evade detection at airports and government buildings. Patrick thinks security is generally lax, and he has himself carried dummy "germ containers" and spray bottles through various security checkpoints just to prove his point.

William C. Patrick, III

Profile

Doctors Without Borders: Bringing Medicine to the Global Village

Kathleen Mahoney, a twenty-five-year-old nurse from Wellesley, Massachusetts, travels to work by small plane, canoe, horse, or foot, using whatever means are available to reach the forty villages in the Amazon region of Brazil that depend on her for medical care. She treats patients with pneumonia, hepatitis, skin and urinary tract infections, influenza, and insect-borne parasitic diseases such as leishmaniasis and malaria. Malaria is endemic to the area, and part of her mission is to train people in the remote villages to diagnose and treat the disease.

Since 1993, microscopes have been provided to nearly all of the villages, and one or two villagers have been trained in their use. As a result, the number of annual cases of malaria in the region has been cut almost in half.

Nurse Mahoney tells about one particular experience: "Recently we hiked one and a half hours through the mountains to see a patient. She was seven months pregnant and had a severe form of malaria. The local health worker and the microscopist, whom we had trained, were concerned about whether they were using the right medicine to treat her. It turned out the microscopist had done the test correctly, and the local health worker had treated her correctly. That was really rewarding. I gave her IV fluids and observed her, and I was confident that she would do all right. We probably won't be back to that village again for three months."

Brigg Reilly, a twenty-nine-year-old epidemiologist from Manhattan Beach, California, has chased infectious diseases in two of the four corners of the globe. After a stint in Zaire, where he worked to prevent a cholera epidemic among Rwandan refugees, his next assignment was in Moscow, where he is coordinating an AIDS prevention project. Moscow is seeing an explosion of HIV infection, with twice as many people testing positive in 1996 as in the previous five years. The group at highest risk is teens, and the route of infection is largely through sharing of needles to inject a home-brewed poppy grass extract with the street name of *vint*.

According to Mr. Reilly, "Russia has long been protected from AIDS by virtue of its isolation, but the disease has finally arrived and threatens to take on epidemic proportions." He hopes that by targeting teens for prevention, "we can take what other countries have learned from the epidemic and apply it to a country that still has a second chance."

Lael Conway, M.D., a thirty-two-year-old internist from Anchorage, Alaska, spent seven months in the island nation of Sri Lanka off the coast of India. It is an area torn by civil war, in which the nation's health care system has collapsed. She worked in mobile clinics set up in villages cut off from larger towns by fighting. Seeing between 50 and 250 patients each day, she treated everything from malaria and malnutrition to rabies, respiratory infections, and infected wounds.

What are these young health care professionals doing in such far-flung areas of the world? Along with Andrew Schechtman, M.D., who studied public health and tropical medicine at Tulane University in New Orleans and has treated and tracked sleeping sickness in Omugo, Uganda; Boston pediatrician Peter Cardiello, M.D., who battled malaria in Myanmar; emergency room nurse Teela Swanson, R.N., from Normal, Illinois, who worked in refugee camps along the Thai/Myanmar border; Viviane Rennard, a nurse-midwife from Albuquerque, New Mexico, who was stationed in Somalia; Anamaria Bulatovic, M.D., a Baltimore native who served in Tanzania and Thailand; New York doctors Bob Zimmerman and Karen Eigen, St. Louis nurse Jennifer Vago, and more than 2,000 other volunteers from more than forty-five nations, they are members of Doctors Without Borders.

Founded in 1971 in France, where it is called Médecins Sans Frontières, this independent, nonprofit international humanitarian organization delivers medical relief anywhere in the world where war, epidemics, or natural disasters threaten the population. Doctors Without Borders currently has missions in more than eighty countries around the globe, from Afghanistan to Uganda and many spots in between.

Volunteers are not paid salaries, though they are given a travel allowance and small monthly stipend while working in the field. In addition to doctors and nurses, members include non-medical personnel to deal with the complex logistics of field operations: communications (including radio links and satellite dishes), transportation (including all-terrain vehicles, airplanes, and boats), construction (including emergency housing and sanitary facilities), and more.

When disaster strikes, Doctors Without Borders first dispatches an exploratory team to assess the situation and report back on the type and quantity of personnel and equipment that will be needed. The challenge is always to get in quickly with teams equipped to deal with the specific situation. During the cholera epidemic that struck seven countries in East and Central Africa in 1997–98, for example, they were able to set up treatment centers within forty-eight hours of the first signs of the epidemic.

One of the ways Doctors Without Borders manages to do this is with a ready stockpile of specialized kits. But if you're thinking about a first-aid kit similar to one you might have at home or in your car, think again. There are fifty different types of kits, some of which include enough medical supplies to treat 10,000 cases of a specific disease for a period of three months. Others might contain generators and other energy sources, motor vehicles, satellite communications equipment, or office supplies. Sanitation kits include 15,000-liter water tanks and equipment for building latrines, digging wells, and purifying water.

Doctors Without Borders does more than respond to emergencies. They also work with local authorities to train medical personnel and establish functioning hospitals and clinics to provide ongoing health care after the Doctors Without Borders team has left.

Currently on the Doctors Without Borders agenda are the needs of refugees and others displaced by war and natural disasters in Africa and Eastern Europe; reemerging infectious diseases such as sleeping sickness, polio, and tuberculosis; new epidemics, including HIV/AIDS; and the collapse of health care systems, such as is happening throughout the former Soviet Union.

The organization's very first mission, back in 1972, was in the Central American nation of Nicaragua, in the aftermath of a devastating earthquake. Two years later, the first Doctors Without Borders long-term medical assistance mission was established in Honduras, which had been nearly wiped off the map by Hurricane Fifi. In the fall of 1998, Doctors Without Borders was back in Central America, lending a hand after Hurricane Mitch. Emergency response teams brought in food, water, and sanitary supplies, as well as medical equipment. Top priority is an offensive against the threat of cholera: workers are building latrines, and installing pumps and generators to reconstruct a basic sanitation system, all the while investigating any suspected outbreaks of the disease.

Where will Doctors Without Borders turn up next? Anywhere in the world where emergency medical care and expertise are needed to help populations at risk because of war or natural disaster.

Healthcare professionals working without borders.

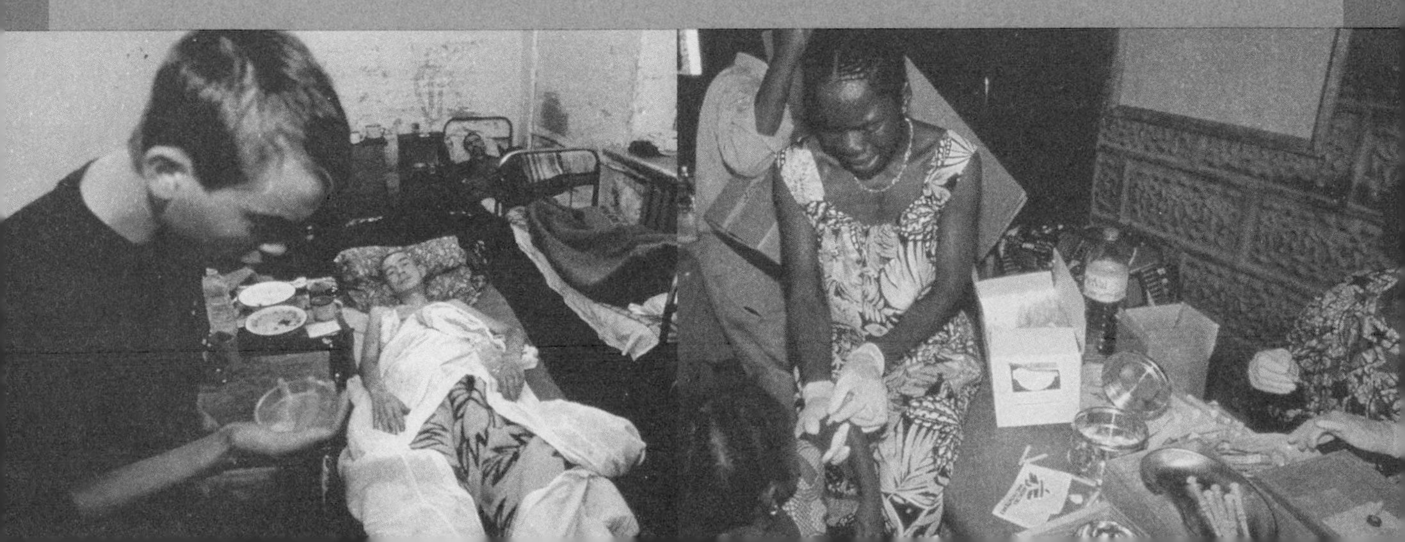

What is WHO and why has it set its sights on malaria?

WHO stands for World Health Organization, an agency of the United Nations committed to public health on a global scale. Founded in 1948, it assists governments in strengthening their health services; tracks epidemics and compiles health statistics worldwide; provides information and assistance in the fight against epidemics, endemics, and other diseases; promotes improved nutrition, housing, sanitation, working conditions, and environmental hygiene; sponsors and conducts research; develops international standards for food, biological, and pharmaceutical products; and provides information and education for the public on health and disease.

That's a huge assignment. It is carried out by doctors, researchers, and other medical and technical personnel from 166 member nations. Six regional offices are in charge of operations in Africa, the Americas, the eastern Mediterranean, Europe, Southeast Asia, and the western Pacific.

The current director-general of WHO is Gro Harlem Brundtland, M.D. She holds a degree in public health from Harvard and served for ten years as the Prime Minister of Norway. When she joined WHO in 1998, one of her first acts was to restructure the organization into clusters, each concerned with an aspect of the overall mission. The task of the Cluster on Communicable Diseases is "to reduce the global impact of communicable diseases through prevention and control, surveillance, and research and development."

In November 1998, Dr. Brundtland announced an ambitious new project partnering WHO and the Cluster on Communicable Diseases with two other UN agencies—UNICEF and the United Nations Development Program—and the World Bank. It is called Roll Back Malaria, and it targets a disease WHO battled once before.

The word malaria is Italian for bad air, because it was thought to be caused by inhaling swamp vapors. Although it is true that swamps and other wetlands are an important factor in the ecology of the disease, malaria is actually caused by one of several protozoan species of the genus *Plasmodium*. *Plasmodium* is spread to humans by the bite of the female *Anopheles* mosquito, which breeds in slow-moving, clear water.

The symptoms of malaria are fever and shaking chills, headache, and joint pain. During part of its life cycle, *Plasmodium* invades the red blood cells, using them as a place to reproduce and destroying the cells in the process. The result is severe anemia and, often, fatal damage to the brain and other vital organs.

Malaria has afflicted people since ancient times, and medicines to treat it are known in many cultures. The Chinese have used an herbal treatment called qinghao for at least 2,000 years, and in Peru the bark of the cinchona tree was found to be effective against malarial fever. Cinchona is also the source of quinine, the standard cure for malaria for more than a century. It was not until the late nineteenth century, however, that the malaria parasite and the means of transmission were identified.

In 1955, WHO initiated a program to eradicate malaria, which involved the use of chloroquine, a synthetic quinine-based drug, and widespread spraying with the insecticide DDT. Within fifteen years, malaria had been driven from most of North America, Southern Europe, the former Soviet Union, and many areas of Asia and South America. But the program backfired in a particularly cruel way.

At the same time as *Plasmodium* began to develop resistance to chloroquine, *Anopheles* mosquitoes were becoming resistant to DDT. Malaria has reemerged as a deadly infection in many areas of the world.

Today, malaria is found mostly in tropical regions of Africa, Asia, and Latin America, and especially where socioeconomic conditions are poor and health systems inadequate. An estimated 300 million people are infected with malaria, and between 1 and 1.5 million die every year. Most of the victims are children, who die of acute infection before they are able to develop immunity. According to WHO, a child dies of malaria every thirty seconds.

The goal of Roll Back Malaria is to cut in half the number of malaria deaths by the year 2010. What will it take to reach this goal?

According to Dr. Brundtland:

- a global strategy to increase speedy access to effective treatment and protection from mosquito bites in areas where the disease is endemic
- intensified efforts and incentives to develop new drugs and insecticides that will be effective against the pathogen and its vector
- economic and technical support to malaria-ridden countries to strengthen their national health systems.

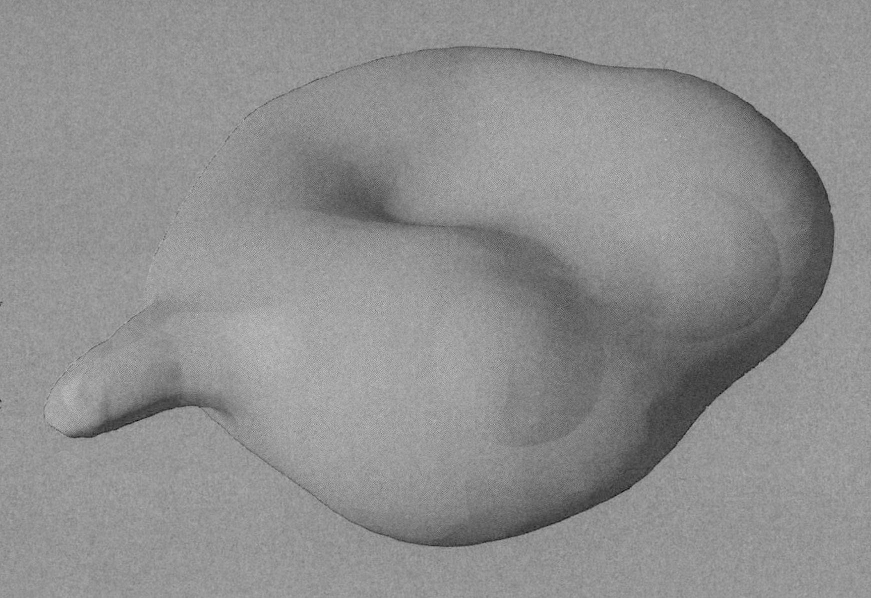

above Computer-generated image of the ringform of the malaria parasite (*Plasmodium falciparum*).

below Malaria vector in Africa, the *Anopheles gambiae* mosquito.

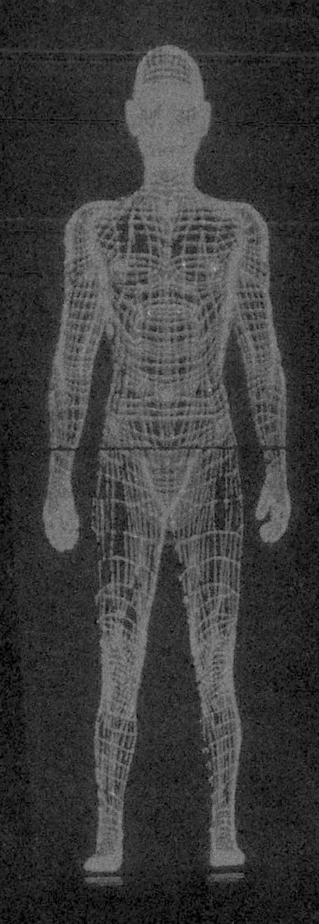

An interactive life-sized computer image shows microbes in the normal body.

About the Exhibition: Epidemic!

The World of Infectious Disease

The exhibition "Epidemic! The World of Infectious Disease," (February to September 1999), continues a long tradition of public health education at the American Museum of Natural History. After leaving New York, the exhibition embarks on a three-year tour to museums of natural history and science throughout the United States. With astonishing models, dioramas, and electronic displays, the exhibition explores the evolutionary challenge of human coexistence with infectious microbes. It shows the immense variety of bacteria, viruses, protozoa, and other microbes, and how we are exposed to them. It illustrates the pathways to infection and how microbes spread through populations, sometimes exploding into epidemics and pandemics.

Disease is a universal global phenomenon that we encounter in the most intimate and individual ways. At the beginning of this century, urban areas such as New York City suffered the scourge of tuberculosis (also called consumption). In 1908, the Museum opened a six-week exhibition, "To Stamp Out the Plague Consumption", that explained the disease and demonstrated its treatment and prevention. The public response was overwhelming. Nearly a million people came out in the winter of 1908–09 to see the exhibit. The museum hired extra security and stayed open until 11 each night to accommodate the crowds.

In the aftermath of this hugely popular exhibition, Dr. Charles-Edward Amory Winslow was appointed to the task of developing a full-fledged department of public health with research, collection, and public education activities. One of his first projects was a 1910 exhibition on New York City Sewers. Winslow's work in the early decades of this century showed that human transformation of the environment disrupts the balance in nature and our relationship with other creatures, including infectious microbes.

By 1914, Winslow had overseen the opening of a permanent Hall of Public Health and the initiation of a public lecture series. Throughout Winslow's tenure, exhibitions reflected the state of scientific knowledge and public concerns of the day. Our water and sewage systems, where we get our food, how we expand and crowd our cities, all have reverberations in the balance of nature that impact human health. Public health education, including the prevention, diagnosis, and treatment of disease, are crucial aspects to restoring and maintaining a healthy environment. Winslow's Hall also featured larger-than-life-size models of the malaria mosquito, the fly, the flea, the louse, and the yellow fever mosquito, some of the common disease vectors. In addition, during the darkest days of World War I, the Museum hosted an exhibition on military hygiene that focused on trench foot and influenza. The Spanish influenza pandemic of 1918–19, spread by troop movements during the war, killed more people than did four years of destructive and bloody combat.

above Models showing how the immune system works. *below* Model of viruses (right) and bacteria (*left*) in the Exposure section.

PUBLIC HEALTH IS AN ISSUE OF NATURAL HISTORY

Since the tuberculosis exhibition at the beginning of the twentieth century, the Museum has frequently explored themes of disease and public health in its exhibitions: "Can Man Survive?" (1969), "New York on Tap" (1986), and "In Time of Plague" (1988). In the Museum's magnificent new permanent Hall of Biodiversity, opened in 1998, many issues of public health are discussed in the context of global biodiversity. On the eve of the twenty-first century, "Epidemic! The World of Infectious Disease" explores the evolutionary history of infectious disease.

Winslow recognized over ninety years ago, that humans are simply part of a larger picture. This insight is a central theme of "Epidemic! The World of Infectious Disease". We are here with all the other organisms on the planet, including pathogenic organisms that can cause us so much grief. The natural history of all the species on the planet unfolds an intricate web of life and death. By examining the coexistence of microbes and humans, we can explore evolution, ecology, and culture.

PROLOGUE

We are all familiar with the sometimes threatening and scary stories of disease outbreaks that we see on the news and in movies. Suddenly a new disease appears—swine flu, Legionnaires', Ebola, AIDS. The media reports suffering and casualties and the efforts of medical workers to bring it under control. The first display in the exhibition surrounds the visitor with brief clips of familiar media coverage. It sets the stage for the exhibition to unravel the natural history of humans and microbes that is key to understanding the often mysterious outbreaks of disease.

On the next step of the journey through the exhibition, the visitor enters a diorama of a toolshed set in the Four Corners region of America's Southwest. In 1993, a disease outbreak here was caused by the hantavirus. A projected image shows someone sweeping the floor of the toolshed, kicking up dust filled with mouse feces, a carrier of the virus. The exhibit begins to detail the evolutionary challenges that interacting species impose on one another.

THE STRUGGLE FOR COEXISTENCE

Humans have evolved along with the microbes that infect us. The exhibition next reveals the long and intensely interactive relationship between humans, mosquitoes, and the malaria protists they carry. The intricate relationship is mediated through the ecology of our common ecosystems. Large, dramatic scrims illustrate the transformation of agricultural villages through controlled flooding and irrigation for rice cultivation, which creates pools of water needed for mosquito habitats. Human efforts to ward off malaria have often been unsuccessful because they did not consider microbial mutation. This example shows how an understanding of evolution can impact our ability to fight disease. In addition, by describing the evolutionary interrelationships between organisms and the environment, the exhibition makes clear that not all microbes are "the enemy." Some play a key role in the delicate balance of life and are essential for human survival.

EXPOSURE

Armed with information about the history and nature of our common life on the planet, the visitor enters the arena where humans and microbes meet: exposure. A microbial gallery filled with approximately 300 large three-dimensional models of bacteria, viruses, protozoa, worms, and fungi fill the space. Text panels open up the world of the microscopic creatures we never see, but which have an impact on our lives so profoundly. Microbes are not simply a single class of organisms. Through the models and explanations, the exhibit portrays the awe-inspiring diversity of the microbial world. The exposure of humans to microbes is complicated by the vast diversity of microbial reproduction, ecology, and relatedness, and is influenced by change in the ecosystems that define microbial human interaction. The exhibit explains the differences between bacteria, viruses, protozoa, helminths, and fungi, according to their structure, reproduction, and evolutionary relationships; and details the "how, when, where, and why" of our encounters with microbes.

In addition to the large models, the exhibition provides a peek at how laboratories work. A variety of laboratory equipment is on display, including an array of microscopes and slides to see what a variety of microbes look

above 12-foot diorama of *E. Coli* O157:H7 in the stomach lining. *below* Model of T-helper cell being infected with the HIV virus.

like; heavy-duty, two eyepiece wencescopes used in medical diagnosis; and a transmission electron microscope to see microbes too small to be visible in a standard light microscope.

INFECTION

How do infectious microbes actually get into our bodies? Once inside, how does our body fight microbial invaders? In this section, the exhibition examines the human body and its defenses against infection.

Humans have also developed several intervention methods for fending off infections, including antibiotics, vaccines, prophylactic and cultural strategies. How do microbes circumvent these elaborate defense systems that our bodies have erected against them?

Microbes enter the human body through five main entry portals: respiration, blood, skin, ingestion, and sexual contact. In this section of the exhibit remarkable computer animation depicts the entry of a flu virus through respiration. Beginning with the inhalation of an airborne particle, it tracks the virus's journey through the body's upper respiratory system.

Microbes utilize the human body to maximize their evolutionary potential. Two dioramas illustrate how microbes attack once they have gained entry. A twelve-foot diameter model of a human T cell infected with HIV shows the process of the virus's replication. An equally large display models the deadly E. *coli* O157:H7 that attaches to the human intestine and secretes poison.

Microbes have evolved a diverse array of lifestyles to maintain themselves in our bodies. But no less amazing than microbial adaptation is the human immune system that fights off, contains, or eliminates these harmful invaders. Large diorama models demonstrate how our immune system works. Our immune system is an incredibly efficient, but not invincible, defender against microbes.

OUTBREAK

In order to survive, microbes must exit the human body and go on to infect others. Infectious microbes have evolved four ways of dispersing among human populations, all part of the complex microbe-human ecosystem: via the respiratory system; the blood (through animal and insect bites and needles); fecal material; and sexual contact. An intricately woven twine model of the human body uses beams of colored lights to illustrate the exit pathways of infectious microbes.

There are six ways that microbes can travel from one person to another: through water, food, vectors, aerosols, sex, and blood. The commonplace conduits microbes utilize to pass through human populations are dramatized on a walkway that passes through a series of dioramas. For example, a food delivery truck and a dinner table illustrate food as a source of microbial transfer. Historical accounts of outbreaks can be useful to understand the history of diseases. In 1854, a devastating outbreak of cholera in London was traced to a single Broad Street water pump. A recreation of this infamous pump reminds us that our water is a medium of microbial transfer. The air we breathe is also a crucial transport for the microbes that live with us. A diorama of an air-conditioning system illustrates how the outbreak of Legionnaires' disease at a hotel in Philadelphia occurred two decades ago. The bow of a ship represents how travel can bring animal and insect carriers that have spread so many diseases through human populations. The reality that our blood is a passage for infection is brought to life in a startling manner by plaster casts of arms sharing a needle. And, finally, the silhouette of two people behind a drawn window shade is meant to remind us that sexual contact is yet another part of our lives which brings us into interaction with the vast and diverse world of microbes.

EPIDEMIC AND PANDEMIC

Epidemics and pandemics are the result of the movement of people on a global scale. The rapid growth of world population and the increase of movement spawned by urbanization, trade, travel, continued war and civil strife, and uneven economic development are all critical factors in the outbreak of epidemics. More than ever we live in a global village easily interconnected by the movement of people and microbes. Even global climate change, perhaps influenced by human economic and cultural behavior, can be an impetus to the spread of disease. A series of video screens display both modern and historical examples of epidemics and pandemics brought about by each of these factors.

malaria protozoan

ACTION

Infectious disease is about the global coexistence and interaction of humans and microbes, which sometimes has dramatic consequences for public health. It touches each of us in the most personal way. All humans are interconnected by the memories they have of infectious disease affecting loved ones and friends. Near the end of the exhibition, visitors enter a sound theater with familiar and evocative objects: a doorway to a clinic, a newsstand, a park bench, scaffolding, a subway platform. "Voices from the Pandemic" lets them listen to the world of HIV/AIDS. They stand in familiar settings and hear personal accounts of how the disease has affected individual lives.

All cultures care for the sick and diseased in their own way. But the essential human commitment is the same. We have a global responsibility to implement care for those inflicted with infectious disease, and informing ourselves about the nature and means of treatment and prevention is essential to that responsibility.

The exhibition ends by emphasizing that personal actions can be incredibly important in our interactions with infectious disease. An Action Area equipped with computer interactives and public health information implements this by connecting the visitor to the World Wide Web and to other resources that provide an opportunity to think about, and even begin, taking action to learn about and fight infectious disease.

It is through knowledge that we gain respect and assume responsibility for our place in the world of infectious disease. It is this sense of respect for natural history and responsibility for our actions that will allow us to coexist on this planet with microbes.

facing page (top) models of red blood cells infected with *Plasmodium falciparum*. *facing page (bottom)* Model of viruses
below View as you enter the Action Resource Center.

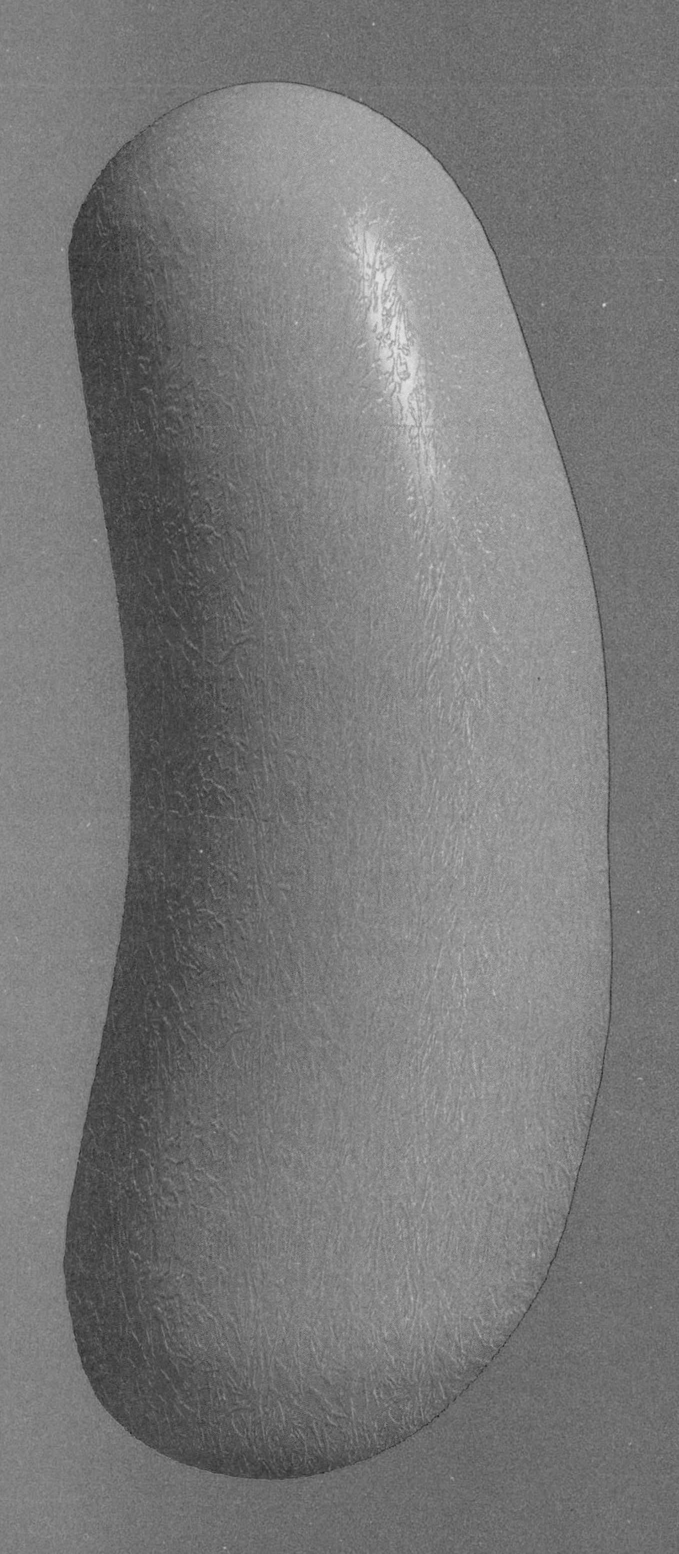

Computer-generated image of the gametocyte form of malaria parasite (*Plasmodium falciparum*).

Resources

Suggested Reading

Alcamo, I. Edward and Lawrence M. Elson. *The Microbiology Coloring Book.* New York: HaperCollins College Publishers, 1996.
Through the use of illustrations, this book introduces the structure, physiology, biochemistry, cultivation, and replication patterns of microorganisms.

Biddle, Wayne. A *Field Guide to Germs.* New York: Henry Holt and Company, 1995.
In a witty account, Biddle introduces nearly 100 infamous pathogens, describing their physical characteristics, symptoms, and treatments.

de Kruif, Paul. *The Microbe Hunters.* San Diego: Harcourt Brace Jovanovich, Publishers, 1926.
From Leeuwenhoek to Paul Erlich, Paul de Kruif dramatizes the work of the scientists whose biological discoveries contributed to the field of microbial research.

Garrett, Laurie. *The Coming Plague: Newly Emerging Diseases in a World Out of Balance.* New York: Farrar, Straus and Giroux, 1994.
In her extensive examination of infectious disease, science writer Laurie Garrett counters the belief that the threat of microbial infection has been effectively eradicated. She highlights contemporary conditions that favor the spread of infectious disease and points to preventative measures that can reduce the risk of emerging diseases.

Giblin, James Cross. *When Plague Strikes: The Black Death, Smallpox,* AIDS. New York: HaperCollins, 1995.
This study of the social, political, and cultural reactions to disease epidemics shows how medical knowledge and treatments have advanced as a result of disease crises.

Karlen, Arno. *Man and Microbes: Disease and Plagues in History and Modern Times.* New York: Touchstone Books, 1996.
Medical historian Arno Karlen writes on the natural history of disease and the coevolution of humans and microorganisms, placing recent outbreaks of deadly diseases in a historical perspective.

Lilienfeld, Abraham M. and David E. Lilienfeld. *Foundations of Epidemiology, 2nd Edition.* New York: Oxford University Press, 1980.
This introductory reference text presents the concepts and methods of epidemiology, emphasizing the integration of biological and statistical elements.

Reese, Richard E., M.D. and Robert F. Betts, M.D., Editors. A *Practical Approach to Infectious Diseases.* New York: Little Brown & Company, 1996.
This desktop reference book on infectious disease includes current views on HIV, new antibiotics, and other up-to-date information on epidemiology.

Roueché, Berton. *The Medical Detectives.* New York: Truman Talley Books, 1988.
Journalist Berton Roueché chronicles true and fascinating medical mysteries and reveals how local health authorities and epidemiologists worked to solve the cases.

Salyers, Abigail and Dixie Whitt. *Bacterial Pathogenesis: A Molecular Approach.* Washington, D.C.: American Society of Microbiologists Press, 1994.
This book gives an extensive introduction to the molecular basis of infectious diseases.

Yancey, Diane. *The Hunt for Hidden Killers: Ten Cases of Medical Mystery.* Brookfield, CT: Millbrook Press, 1994.
Yancey describes how doctors and public health investigators work to solve the enigmas surrounding odd symptoms, unexplained poisonings, and outbreaks of rare or previously unknown diseases.

Videos, Films, and Other Media

Epidemic!: A Fred Friendly Seminar
PBS Video
1320 Bradddock Place
Alexandria, VA 22314-1698
1-800-334-3337
http://shop.pbs.org
The Fred Friendly Seminar uses Socratic dialog to promote discussion about the politics, economics, and science of infectious disease.

National Science Teachers Association (NSTA).
The Science of HIV.
To order, call 1-800-722-NSTA
This teacher's guide and video provide learning opportunities ranging from labs, activities, readings, model design, and guided discussions.

Human Relations Media. Viruses: The Deadly Enemy.
To order, call 1-800-431-2050
This video covers the fundamentals about disease, the immune system, evolution, and genetics and uses graphic animation to demonstrate how viruses are structured.

Web Sites

The AIDS Page for Teens
http://hyperion.advanced.org/10631/index.html
Specifically geared toward teens, this site provides information about AIDS prevention.

Alliance for Prudent Use of Antibiotics (APUA)
www.healthsci.tufts.edu/APUA/apua.html
The APUA site provides information on antibiotic resistance and includes an online newsletter.

The American Museum of Natural History
www.amnh.org
The American Museum of Natural History's site highlights the Museum's activities, research, and educational outreach programs.

Australian Society for Microbiology (ASM)
http://home.vicnet.net.au/~asm/welcom2.htm
ASM is a professional society dedicated to furthering the science of microbiology. The Web site offers information about ASM and links to related sites.

Cells Alive!
www.cellsalive.com
Cells Alive! offers a peek into a vast image library of microbes. It includes a helpful section that identifies online resources for specific science topics.

The Centers for Disease Control and Prevention
www.cdc.gov/ncidod/ncid.htm
The Centers for Disease Control and Prevention site provides access to current disease statistics, and health precautions for travelers, as well as general health information.

Digital Learning for Microbial Ecology
http://commtechlab.msu.edu/sites
The Digital Learning Center for Microbial Ecology (DLC-ME) is a science education project under development at Michigan State University. The site features a "Microbe of the Week" and "The Microbe Zoo," a virtual space where visitors can view images and descriptions of microscopic organisms and the habitats in which they live.

Doctors Without Borders
www.dwb.org/index.htm
The site gives general background and regular updates on the activities of Doctors Without Borders. The site also features a disease directory and volunteer information.

HIV InSight
http://hivinsite.ucsf.edu
An AIDS resource site, HIV InSight is a project of the University of San Francisco AIDS Program at San Francisco General Hospital and the UCSF Center for AIDS Prevention Studies. The site offers medical information, news, and resource links.

Pan American Health Organization (PAHO)
http://www.paho.org
This site offers public information about the Pan American Health Organization in both English and Spanish. Visitors can find country health profiles, a catalog of publications,and information about PAHO's Library and Information Services.

The Wellcome Trust Centre for the Epidemiology of Infectious Disease
http://www.ceid.ox.ac.uk/
In addition to general information about the Centre, the site has links to the Centre's interdisciplinary research topics in epidemiology, population biology, evolution, and control of infectious disease.

Scientific American
www.scientificamerican.com
Scientific American's Web site features scientific news. Articles on infectious disease appear frequently.

The World Health Organization (WHO)
www.who.org
The World Health Organization's site provides access to information on health topics, including disease, environment and lifestyle, and health policies.

Exterior of the American Museum of Natural History, circa 1900.

About the American Museum of Natural History

Founded in 1869, the American Museum of Natural History is one of the world's preeminent institutions for scientific research and education.

Today, under the direction of President Ellen V. Futter, the Museum's scientific, education, and exhibition staff are working to discover, interpret, and disseminate knowledge and human cultures, the natural world, and the universe. Prepared for the challenges of twenty-first century society, the Museum is committed not only to making contributions to science, but to improving science education and enhancing science literacy nationwide. In 1997, the Museum launched the National Center for Science Literacy, Education, and Technology to take the Museum beyond its walls to a national audience. The National Center uses media and technology to connect people of all ages to real scientists and their work. The purpose of the National Center is to take the Museum's vast resources—collections of some thirty-two million specimens and artifacts, forty-three exhibition halls, and a staff of more than 200 scientists, and 128 years of expertise in educational programming—directly to classrooms, libraries, community centers, and homes throughout the country.

List of Contributors

Dr. Jeremiah A. Barondess was educated at the Pennsylvania State University and the University of Michigan and received his M.D. from Johns Hopkins University. Dr. Barondess' particular clinical interests have been in diagnosis and the diagnostic process. He has written extensively on clinical matters in internal medicine, on medical education, clinical ethics, and the training of internists. Since Dr. Barondess's appointment to the Presidency of The New York Academy of Medicine in July 1990, he has directed its programmatic activities primarily in the areas of urban health, recruiting to the health professions and medical education, the medicine/science/society interface and the health of the biomedical enterprise itself.

Dr. A. David Brandling-Bennett received his undergraduate and medical degrees from Harvard University and a diploma in tropical public health from the London School of Hygiene and Tropical Medicine. As an epidemiologist with the Centers for Disease Control and Prevention, he worked in Central America, Thailand, and Kenya. He joined PAHO in 1989 as Head of Epidemiology, later becoming Director of Communicable Diseases. He was appointed PAHO's Deputy Director in 1995.

Dr. Marla Jo Brickman received her doctorate in microbiology and immunology from Duke University. Her research interests are in cell biology and parasitology. Dr. Brickman received an American Association for the Advancement of Science Mass Media Science and Engineering Fellowship. She spent her fellowship at the *Atlanta Journal-Constitution* as a health and science reporter. Dr. Brickman also worked as a public affairs officer at Memorial Sloan-Kettering Cancer center. She has written numerous scientific articles.

Dr. Barry R. Bloom is the Dean of the Harvard School of Public Health and Professor in the Department of Immunology and Infectious Disease. He is an immunologist and microbiologist engaged in research on infectious diseases and vaccines. Dr. Bloom chairs the Vaccine Advisory Committee of UNAIDS. He serves as a member of the U.S. AIDS Vaccine Research Committee, the National Academy of Sciences, the American Academy of Arts and Sciences, and as counsellor of the Center for Infectious Diseases at the CDC. He served on the Advisory Council of the National Institute of Allergy and Infectious Disease.

Dr. John Caldwell and Pat Caldwell have researched African demography and health for almost forty years. They have also carried out research in South and Southeast Asia. They direct a collaborative program of research on the cultural, social, and behavioral context of AIDS in Africa involving the Australian National University and universities in Uganda, Ghana and Nigeria, funded by the Swedish International Agency. John Caldwell was head of the Australian National University's Department of Demography for twenty years, is an Adjunct Professor of the Harvard School of Public Health, and is the author of twenty books and over 200 papers.

Dr. Richard J. Colonno is currently Vice President of Infectious Diseases Drug Discovery at Bristol-Myers Squibb Co. in Wallingford, Connecticut, where he is responsible for the discovery and early development of novel antiviral, antibacterial, and antifungal agents. He is a molecular virologist by training who has been involved in antiviral research for almost twenty years. Dr. Colonno is one of the world's leading experts on Human Rhinoviruses (common cold virus) and has authored nearly 100 original scientific articles on a wide variety of different viruses, including HIV, hepatitis B virus, influenza virus and herpes viruses. He and his team have advanced several early-stage inhibitors into clinical trials.

Dr. Rob DeSalle received his B.A. in Biology at the University of Chicago and his Ph.D. in Ecology and Evolutionary Biology at Washington University in St. Louis. He was appointed an Assistant Professor in the Department of Biology at Yale University in 1986 and moved to a curatorial position at the American Museum of Natural History in 1990, where he now co-directs the Molecular Laboratories. He is the author of *The Science in Jurassic Park and The Lost World* (Basic Books, 1997) and curated "Epidemic! The World of Infectious Disease."

Laurie Garrett did graduate work in immunology before embarking on a career in science journalism. For eight years she was science correspondent for National Public Radio, and for the last ten she has covered the beat for *Newsday*. Garrett, who has won the Pulitzer, Polk, and Peabody Awards, was a fellow at the Harvard School of Public Health in 1992–93, where she wrote *The Coming Plague: Newly Emerging Diseases in a World Out of Balance*.

Dr. Anne A. Gershon has had a thirty year interest in varicella and zoster. She has published widely on the subject, including studies of the natural history of disease caused by VZV, diagnostic procedures, safety and efficacy studies of vaccination of leukemic children and healthy adults, and helped to develop recommendations for vaccine use. She has been a member of several national committees dealing with recommendations for use of vaccines. She is the Chief Scientific Advisor to the VZV Research Foundation in New York City.

Dr. Gregory E. Glass is an Associate Professor in the Department of Molecular Microbiology and Immunology at the Johns Hopkins School of Hygiene and Public Health and directs the GIS Unit of the Program on Health Effects of Global Environmental Change. He obtained his Ph.D. in Systematics and Ecology from the University of Kansas, and post-doctural training at Johns Hopkins School of Public Health. He has published widely in the field of satellite remote sensing and geographic information systems (GIS) as applied to public health, especially for the vector-borne diseases.

Dr. Paul Greenough is a Professor of History at the University of Iowa and teaches the history of public health. He published a study entitled *Prosperity and Misery in Modern Bengal: the Famine of* 1943–44 (Oxford University Press, 1982), and has since then conducted research on immunization, social welfare, environmental history, health, and local politics in India. He serves as the Director of the Global Health Studies Program at the University of Iowa and has been an associate editor of the *Journal of Asian Studies*. At present he is completing a history of the Epidemic Intelligence Service of the U.S. Centers for Disease Control and Prevention.

Dr. Francesca T. Grifo is Director of the Center for Biodiversity and Conservation at the American Museum of Natural History. Her interests center on the conservation of biodiversity, including how scientific results are best integrated into conservation projects, policy, and education. She currently oversees projects that demonstrate how this integration is possible. She has focused on intellectual property rights and benefits-sharing issues related to the commercialization of biodiversity, including how these and other issues relevant to scientists are interpreted through the Convention on Biological Diversity. Additionally, she has worked closely with an array of institutions in Eastern Europe on national-level biodiversity management and planning. Her recent work has examined the relationships between biodiversity and human health. She holds an adjunct appointment at Columbia University.

Dr. Margaret A. Hamburg, has a distinguished record of scientific accomplishments and outstanding public service. As the Assistant Secretary for Planning and Evaluation at the U.S. Department of Health and Human Services, Dr. Hamburg serves as the principal policy advisor to the Secretary of Health and Human Services. Prior to her current appointment, Dr. Hamburg served as the Commissioner of Health for the City of New York. As Chief Health Officer for the nation's largest city, her many accomplishments included the design and implementation of an internally recognized tuberculosis control program that produced dramatic declines in tuberculosis cases and the development of the most comprehensive local bioterrorism preparedness planning effort in the nation. Dr. Hamburg also served as

Assistant Director of the National Institute of Allergy and Infectious Diseases, National Institutes of Health, where she played a major role in research, administration, and policy development. Dr. Hamburg is a graduate of Harvard College and Harvard Medical School.

Dr. Varuni Kulasekera received her doctorate in Entomology from University of Maryland, College Park. Her research interests are in Systematics and Evolution. She is studying the evolution of flies, with emphasis on disease vectors, at the American Museum of Natural History. She has received fellowships from the National Science Foundation and the American Museum of Natural History to study the evolution of flies.

Dr. Joshua Lederberg was educated at Columbia and Yale University where he was a pioneer in the field of bacterial genetics. In 1958, Dr. Lederberg received the Nobel Prize in Physiology or Medicine for his work and subsequent research on bacterial genetics. He has been a professor of genetics at the University of Wisconsin and at Stanford University School of Medicine. He retired as president of the Rockefeller University in July 1990 and continues his research activities there in the field of DNA secondary structure and mutagenesis in bacteria. He recently co-chaired the study of Emerging Infections which has appeared under the imprint of the Institute of Medicine. Dr. Lederberg has a long-standing interest in the communication of science to the public and for public policy. He authored a weekly syndicated newspaper column for six years and he co-chaired the Carnegie Commission on Science, Technology and Government.

Dr. Joseph E. McDade, the Deputy Director of the National Center for Infectious Diseases, Centers for Disease Control and Prevention, has a long-standing interest in the subject of new and reemerging infectious diseases. He is editor of the journal *Emerging Infectious Diseases*, which tracks infectious disease trends and analyzes infectious diseases around the world. The journal can be accessed via CDC's home page at http://www.cdc.gov.

Anne Platt McGinn is a Research Associate at the World Watch Institute where she researches environmental health, fisheries, and ocean issues. She is a regular contributor to the Institute's annual publications, "Vital Signs" and "State of the World," and wrote World Watch Paper #142, "Rocking the Boat: Conserving Fisheries and Protecting Jobs," and World Watch Paper #129, "Infecting Ourselves: How Environmental and Social Disruptions Trigger Disease." She has also published many articles in *World Watch* magazine, including an article on the re-emergence of tuberculosis, which was selected as one of the top ten "censored" news stories of 1994, by Sonoma State University's Project Censored. Prior to joining the Institute in April 1993, she worked as a Research Assistant at the Environmental Law Institute. Ms. McGinn graduated from Williams College in 1991 with a B.A. in political science and a concentration in environmental studies.

Dr. Stephen S. Morse is Director, Program in Emerging Diseases (and Assistant Professor of Epidemiology), Columbia University School of Public Health, and an adjunct faculty member at The Rockefeller University. He is currently serving in Washington as a Program Manager at the Defense Advanced Research Projects Agency (DARPA). He is the editor of two books, *Emerging Viruses* (Oxford University Press, 1993; paperback, 1996), and *The Evolutionary Biology of Viruses* (Raven Press, 1994), and is an editor of the journals *Emerging Infectious Diseases* (published by the CDC) and *Research in Virology* (Pasteur Institute). He chaired the 1989 Conference on Emerging Viruses (National Institutes of Health), was a member of the Committee on Emerging Microbial Threats to Health at the Institute of Medicine-National Academy of Sciences, is currently a member of the Steering Committee of the Institute of Medicine's Forum on Emerging Infections, and is chair of ProMED (the Program for Monitoring Emerging Diseases).

Dr. Alvin Novick's research has concerned ethics, public policy, and community development aspects of the HIV/AIDS epidemic since 1982. He has been particularly interested in the consequences of society's substantial and stigmatizing disdain for a high proportion of those who are vulnerable to HIV infection. His current principal focus is on attempts to modulate the interactions of HIV/AIDS with the equally troubling epidemic of illicit drug use. In addition to his academic posts, Dr. Novick is editor-in-chief of *AIDS & Public Policy Journal.*

Dr. Jonathan A. Patz, is a Research Assistant Professor and Director of the Program on Health Effects of Global Environmental Change at the Johns Hopkins School of Hygiene and Public Health. He has medical boards in both Occupational and Environmental Medicine and in Family Medicine. He obtained his M.D. degree from Case Western Reserve and his Masters in Public Health at Johns Hopkins. He serves as a Principle Lead Author for assessments of the United Nations Intergovernmental Panel on Climate Change (IPCC), is co-chair for the health sector of the U.S. Global Change Research Program (USGCRP) National Assessment on Climate Change, and has been called on to brief both Congress and the Administration on these matters.

Dr. Robert E. Shope and his colleagues at the University of Texas Medical Branch maintain the World Reference Center for Arboviruses. This collection contains over 500 different types of viruses transmitted by mosquitoes, ticks, and other arthropods. He discovered several of these viruses and has studied their natural history and epidemic behavior in Malaysia, Brazil, Egypt, and many other overseas settings over the forty-five years. He taught epidemiology at Yale University School of Medicine for thirty years, and now teaches at the University of Texas Medical Branch at Galveston. He co-chaired with Dr. Lederberg the Institute of Medicine's 1992 study "Emerging Infections: Microbial Threats to Health in the United States."

Dr. Andrew Spielman is the Director of the Malaria Epidemiology Program of the Center for International Development at the Kennedy School of Government and is Acting Director of the Epidemiology of Infectious Disease Program at the Harvard School of Public Health. His research interests focus on the mode of transmission of microbes between their insect or tick carriers and their human or other vertebrate hosts. He established the life cycle of the agent of Lyme disease and discovered how this bacterium perpetuates itself in nature. He determined how the juvenile and growth hormones of insects regulate the life processes of mosquitoes and how sperm is transferred from male to female mosquitoes. His nearly 300 scientific reports deal with various other processes in spiders, ticks, lice, flies, and mosquitoes and with their disease relationships.

Dr. Louis W. Sullivan was born in Atlanta and received a bachelor of science degree, magna cum laude, from Morehouse College and earned his medical degree, cum laude, from Boston University. He was sworn in as Secretary of Health and Human Services (HHS) on March 10, 1989, and served as Secretary throughout President George Bush's administration. Dr. Sullivan went to HHS from Morehouse School of Medicine, an institution at which he played a founding role. In 1975, Dr. Sullivan became the founding dean and director of the Medical Education Program at Morehouse College. In 1981, Dr. Sullivan became the first Dean and President when the School of Medicine became independent from Morehouse College.

Credits

COVER

Computer-generated image of microbes, courtesy of James Stoop and Barrett Klein, American Museum of Natural History.

Workers in BioSafety Level 4 suits, courtesy of the Centers for Disease Control and Prevention.

TITLE PAGE

Computer-generated image of microbe, courtesy of James Stoop and Barrett Klein, American Museum of Natural History.

FOREWARD

American Museum of Natural History—circa 1900, courtesy of the American Museum of Natural History.

SECTION ONE

Louis Pasteur, courtesy of the American Museum of Natural History.

Computer-generated image of microbe, courtesy of James Stoop and Barrett Klein, American Museum of Natural History.

E. Coli 0157, courtesy of the Centers for Disease Control and Prevention.

Serengeti Plain, photo Richard G. Van Gelder, courtesy of the American Museum of Natural History.

Computer-generated image of microbe, courtesy of James Stoop and Barrett Klein, American Museum of Natural History.

"Mutation," illustration by Sean Murtha.

"*Ixodes dammini*, "illustration by Sean Murtha.

Life cycle of ticks, courtesy of the Centers for Disease Control and Prevention.

Computer-generated image of microbe, courtesy of James Stoop and Barrett Klein, American Museum of Natural History.

A hut in western Bolivia, photo courtesy of Francesca T. Grifo.

Forest remnant, Azuay, Ecuador, photo courtesy of Francesca T. Grifo.

Soil erosion in Bolivia, photo courtesy of Francesca T. Grifo.

René Dubos, photo courtesy of The Rockefeller University.

SECTION TWO

Avian influenza, courtesy of C. Goldsmith and J. Katz/Centers for Disease Control and Prevention.

Computer-generated image of microbe, courtesy of James Stoop and Barrett Klein, American Museum of Natural History.

Computer-generated image of microbe, courtesy of James Stoop and Barrett Klein, American Museum of Natural History.

Computer-generated images of *Shigella dysenteriae*, spores of *Histoplasma capsulatum*, *Giardia intestinalis*, hepatitis B, *Mycobacterium tuberculosis*, and *Planaria dugesia*, courtesy of James Stoop and Barrett Klein, American Museum of Natural History.

Prions in "mad cow" brain, electron micrograph: EM Unit, VLA/Science Library/Photo Researchers.

Hantavirus, courtesy of the Centers for Disease Control and Prevention.

Global Microbial Threats in the 1990s, courtesy of the Centers for Disease Control and Prevention.

"Vector/Non-vector Transmission," illustration by Sean Murtha.

Map of Broad Street, London, courtesy of the Wellcome Institute Library, London.

Smallpox, courtesy of C.S. Goldsmith and J. J. Esposito/Centers for Disease Control and Prevention.

Child being vaccinated in the United States, photo courtesy of the Centers for Disease Control and Prevention.

Child being vaccinated in Thailand, photo courtesy of WHO/A. S. Kochar.

Portrait of Mary Wilson, photo courtesy of Mary Wilson.

SECTION THREE

Computer-generated image of microbe, courtesy of James Stoop and Barrett Klein, American Museum of Natural History.

HIV, courtesy of the Centers for Disease Control and Prevention.

Computer-generated image of microbe, courtesy of James Stoop and Barrett Klein, American Museum of Natural History.

Children in Peru, photo Bobbie Person, courtesy of the Centers for Disease Control and Prevention.

Hamburger, photo courtesy of the Centers for Disease Control and Prevention.

"Immune Response," illustration by Sean Murtha.

Child with chicken pox, photo John Cunningham/Visuals Unlimited.

Dr. Anne Gershon with Dr. Philip La Russa, photo courtesy of Dr. Anne Gershon.

Computer-generated image of microbe, courtesy of James Stoop and Barrett Klein, American Museum of Natural History.

Lab photos, courtesy of the Centers for Disease Control and Prevention.

Preparing the flu vaccine, courtesy of the Centers for Disease Control and Prevention.

Portrait of Stuart Levy, courtesy of Stuart Levy.

SECTION FOUR

Smallpox, courtesy of the Centers for Disease Control and Prevention.

Workers in BioSafety Level 4 suits, courtesy of the Centers for Disease Control and Prevention.

Computer-generated image of microbe, courtesy of James Stoop and Barrett Klein, American Museum of Natural History.

"The Boscombe Valley Mystery" in Sir Arthur Conan Doyle, *The Complete Sherlock Holmes* (Garden City N.Y.: Doubleday & Company, 1927 [reprint ed. Undated]), 211.

Basil Rathbone as Sherlock Holmes, AP/Worldwide Photos.

Hog woodcut, 1895, Corbis Bettmann.

"A Pig From Jersey," from *The Medical Detectives* , by Berton Roueché, Truman Talley Books, New York.

Bonnie Smoak in the field, courtesy of Bonnie Smoak.

Legionella, courtesy of the Centers for Disease Control and Prevention.

Legionella on agar plate, photo courtesy of Jim Gathany, Centers for Disease Control and Prevention.

Bellevue-Stratford floorplan, Smithsonian Institution, NMAH/Medical Sciences.

Resident dwelling, courtesy of the Centers for Disease Control and Prevention.

Peromyscus maniculatus, photo L.L. Master, courtesy of the American Society of Mammalogists.

Piñon nuts, photo courtesy of the Centers for Disease Control and Prevention.

Hantavirus, courtesy of the Centers for Disease Control and Prevention.

Hantavirus Risk Map, courtesy of Jonathan Patz.

HIV electron micrograph (EM) 151,000x, courtesy of Alyne Harrison and Paul Feorino/Centers for Disease Control and Prevention.

HIV scanning electron micrograph (SEM) 24,000x, courtesy of C.S. Goldsmith/Centers for Disease Control and Prevention.

SECTION FIVE

Computer-generated image of microbe, courtesy of James Stoop and Barrett Klein, American Museum of Natural History.

Lab workers, photos courtesy of the Centers for Disease Control and Prevention

"Significant Outbreaks of Smallpox in New World Peoples 1507-1634," illustration by Lourdes-marie Prophete. Dates for map from *Princes and Peasants: Smallpox in History* by Donald R. Hopkins, The University of Chicago Press, Chicago.

Water fountain, photo courtesy of the Centers for Disease Control and Prevention

Rio de Janeiro, courtesy of the United Nations/Shelley Rotner.

World TB Day poster, courtesy of the New York City Department of Health, Neal L. Cohen, M.D., Commissioner of Health.

TB examination, courtesy of the American Lung Association.

HIV, courtesy of the Centers for Disease Control and Prevention.

AIDS Belt 1997, courtesy of John Caldwell.

Portrait of Anthony Fauci, courtesy of Anthony Fauci.

SECTION SIX

Computer-generated image of microbe, courtesy of James Stoop and Barrett Klein, American Museum of Natural History.

City bus, photo courtesy of the Centers for Disease Control and Prevention.

Measles, courtesy of the Centers for Disease Control and Prevention.

Needle pack, photo courtesy of the Centers for Disease Control and Prevention.

Polio, courtesy of Joe Esposito/Centers for Disease Control and Prevention.

HIV, courtesy of Alyne Harrison and Paul Feorino/ Centers for Disease Control and Prevention.

AIDS awareness pamplets, photo courtesy of Pan American Health Organization.

Portrait of Mathilde Krim, photo by William G. Harris.

Computer-generated image of microbe, courtesy of James Stoop and Barrett Klein, American Museum of Natural History.

Portrait of William Patrick, courtesy of William Patrick.

Tuberculosis patients in Mariinsk prison in Siberia, Russia, photo Roger Job, courtesy of Doctors Without Borders.

Healthcare worker vaccinates a child during 1996 meningitis outbreak in Nigeria, photo Roger Job, courtesy of Doctors WIthout Borders.

Anopheles gambiae, photo courtesy of Jim Gathany/Centers for Disease Control and Prevention.

Computer rendering of *Plasmodium falciparum*, courtesy of James Stoop and Barrett Klein/American Museum of Natural History.

ABOUT THE EXHIBITION

Photos of the exhibition, Epidemic! The World of Infectious Disease, courtesy of Denis Finnin/American Museum of Natural History.

ABOUT THE MUSEUM OF NATURAL HISTORY

American Museum of Natural History—circa 1900, courtesy of the American Museum of Natural History.

Glossary

AIDS (Acquired Immune Deficiency Syndrome)—a virus that weakens the immune system and subjects the patient to opportunistic diseases such as pneumonia and tuberculosis. The virus is transmitted through the exchange of body fluids, primarily semen, blood, and blood products. Though progress has been made in prolonging the life of AIDS patients, there is currently no cure for the disease. However, it can be prevented by having protected sex and using sterilized needles. AIDS is a problem throughout the world, especially in sub-Saharan Africa.

aerobic—growing or thriving only in the presence of oxygen.

aerosol transmission—a cloud or mist of solid or liquid particles containing pathogenic microorganisms, released by sneezing or coughing.

amoebiasis—disease caused by the protozoan *Entameoba histolytica*, a type of ameoba.

analgesics—drugs that relieve pain.

antibiotic—a substance made either from a mold or bacterium, or synthetically, that inhibits or kills certain microorganisms, specifically bacteria, and that treats infections.

antibodies—any of a large number of proteins of high molecular weight that are produced, as part of the immune response, in response to an antigen which it then neutralizes, tags, or destroys.

antigen—any foreign substance that when introduced into the body, stimulates an immune response.

arthropod—an invertebrate having jointed limbs and a segmented body with an exoskeleton.

asexual reproduction—a type of reproduction where an organism replicates itself, by budding or dividing, without the involvement of other organisms.

attenuated—reduced in strength.

autoimmune disease—any of a large group of diseases characterized by abnormal functioning of the immune system that causes it to produce antibodies against the body's own tissues.

autotrophs—microorganisms that use inorganic materials as sources of nutrients.

B cells—one of the two major classes of lymphocytes; during infections, B cells mature into plasma cells, which produce antibodies directed at specific antigens.

Babesiosis—a rare, often severe (and sometimes fatal) illness that is caused by a pathogen transmitted by ticks.

bacterium—a single-celled microscopic organism, whose genetic material is not enclosed by a membrane.

binary fission—a form of asexual reproduction in which a cell divides into two daughter cells after DNA replication.

biological warfare—the use of bacteria or viruses, or their toxins as weapons.

bubonic plague—a bacterial infection, transmitted from the flea bite of an infected rat to humans. Symptoms include high fever, chills, weakness, and enlarged lymph nodes that turn black (hence the name the Black Death). The plague originated in China and spread to Western Asia and Europe because China was one of the busiest trading nations. This devastating disease killed one third of Europe's population over a five-year period. Though the plague no longer exists, the basic elements of transmission do make future epidemics a possibility. Preventative measures are the proper disposal of garbage and protecting household animals from flea infestation.

budding—a form of asexual reproduction where a bud or outgrowth from the end or side of the parent cell emerges and develops into a new organism.

capsid—the outer protein shell surrounding the nucleic acid of a virus.

capsomeres—repetitive protein subunits which form the capsid; often arranged in a symmetric pattern.

carbohydrate—chemical substances containing carbon, oxygen, and hydrogen atoms which are an essential structural component of living cells and a source of energy for animals (i.e., sugars and starches).

carotid arteries—paired large caliber vessels that pass on either side of the neck supplying oxygenated blood to the brain.

cell-mediated—the branch of the immune system in which specific defense cells, rather than antibodies, respond and act against a foreign antigen.

cellulose—a complex carbohydrate that is the chief element of all plant tissues and fibers.

Centers for Disease Control and Prevention (CDC)—a governmental agency whose mission is to promote health and quality of life by preventing and controlling disease, injury, and disability.

Chagas' disease—a parasitic infection that is transmitted by biting insects; it can be prevented by sleeping with bed netting and using insecticide to kill insects.

chemolithotrophs—organisms which obtain their energy from the oxidation of inorganic compounds.

chemostat—an apparatus designed to grow bacteria indefinitely, while keeping the conditions and the colony size constant by having a continuous flow of liquid nutrient wash the colony and steadily remove bacteria.

chicken pox—a viral infection spread through direct contact or by coughing, sneezing, and touching contaminated clothing causing a blister-like rash on the surface of the skin and mucous membranes. The blisters first appear on the face and then spread to the entire body. Some children may have a few, and others have several hundred. A mild fever or a general malaise are other symptoms that may accompany this infection. In 1995, a vaccine was developed that took fifteen years to receive approval from the Food and Drug Administration, though it is not widely used. If a child does contract the disease, a topical solution is applied to the blisters to soothe the itching. In very rare cases, children have developed serious complications that have proved fatal.

chlorophyll—a group of green pigments found in green plants, algae, and some bacteria necessary for energy production.

cholera—a bacterial infection transmitted by fecal-contaminated food and water and by ingesting raw or undercooked seafood. Symptoms include diarrhea, abdominal cramps, nausea, vomiting, and severe dehydration. Cholera is endemic to India, Africa, the Mediterranean, South and Central America, Mexico, and the United States. The treatment that is administered is usually antibiotics, which shorten the duration of the illness. Death can result from the severe dehydration due to diarrhea. Contaminated water supply is the main mechanism for spreading the disease and improving sanitation conditions serves as the best form of prevention. A vaccination is available as a short-term method to countries at risk. The smuggling of shellfish led to the reemergence of cholera in the United States, so avoiding raw or undercooked seafood would reduce the risk.

chromosome—a threadlike body in the cell nucleus that carries the genes in a linear order.

cilia—short, hair-like appendages found on the surfaces of some types of cells and organisms; used for either propelling trapped material out of the body or for locomotion.

coadaptation—mutual adaptation in two or more interactive species.

computerized axial tomography (CAT scan)—a special radiographic technique that uses a computer to assimilate multiple x-ray images into a two-dimensional, cross-sectional image.

conjugation—a mating process where the temporary union of two one-celled organisms results in the exchange of genetic material.

cryptosporidiosis—an infection caused by an intestinal parasite, transmitted through the ingestion of food or water contaminated with animal feces.

cytoplasm—the living substance of a cell excluding the nucleus.

deforestation—the state of being clear of trees.

dengue fever—a virus transmitted from the bite of the *Aedes* mosquito. Symptoms include headaches, fever, joint pain, and a rash. The disease can be fatal if the patient goes into shock. Dengue fever occurs in most tropical areas, more recently in Asia, the Pacific, and the Americas. The peak feeding activity of this mosquito is after daybreak and in the late afternoon. Wearing protective clothing that covers most of the body, arms and legs, and using insect repellent with the chemical DEET, are ways to prevent the disease. Prevention and control can be accomplished through mosquito control. Eradication of the mosquito with insecticides are the best preventative measures.

diagnosis—the act of identifying a disease and its cause.

dialysis—a medical procedure that uses a machine to filter waste products from the bloodstream and restore the blood's normal constituents.

Diphtheria—an acute bacterial infectious disease that is spread by droplets sprayed from an infected person; children can be immunized against this disease.

DNA (deoxyribonucleic acid)—the primary genetic material of a cell.

Ebola—a deadly virus that is transmitted through direct contact with the blood or bodily fluids of an infected person, unsterilized needles, or an infected animal. Symptoms include high fever, headaches, muscle aches, stomach pain, fatigue, and diarrhea. Initial symptoms occur four to sixteen days after infection, and if not caught in time, proceed very rapidly. Ultimately, the virus causes all major organs to fail, resulting in death. In 1976 in Zaire, the disease appeared from an unknown place and killed 340 people. The origin of the virus was in the Congo, moving to the Sudan, the Ivory Coast, Zaire, and most recently in 1996 in rural Gabon. Early diagnosis is important in treating the virus, and extreme infection control methods such as isolation are the only way to prevent the spread of the virus.

ecosystem—a community of organisms and their physical environment interacting as an ecological unit.

electron microscopy—a form of microscopy in which a beam of electrons deflected by electromagnets can magnify a specimen up to 400,000 times its original size.

Encephalitis—a virus caused by the bite of an infected mosquito; it can be prevented by wearing long pants and long sleeves, staying inside from dusk to dark when mosquitos feed, and using insect repellent.

endemic—a disease that is constantly present to a greater or lesser degree in people living in a particular location.

endoplasmic reticulum—an extensive network of internal membranes within an eukaryotic cell which is necessary for protein synthesis.

enzyme—any of several complex proteins that are produced by cells and act as catalysts in specific biochemical reactions.

epidemic—a widespread outbreak of an infectious disease where many people are infected at the same time.

epidemiology—the branch of medical science dealing with the incidence, distribution, and control of disease in a population.

eukaryote—a cell that possesses a defined nucleus surrounded by a membrane; protists, fungi, plants, and animals are eukaryotes.

extremophiles—organisms (typically bacteria) that are adapted to living in extreme conditions, such as high salt, ice, or thermal springs.

fermentation—the oxidation of compounds by the enzyme action of microorganisms

flagellum—a thin, filamentous appendage on cells, such as bacteria and protists, responsible for locomotion.

flora—in microbiology, the microorganisms present in a given environment; normal flora are those microorganisms which reside harmlessly within the human body.

fluke—a parasitic trematode worm, which has a flat, leaf-shaped body and two suckers.

fungi—molds, mushrooms, and yeasts which comprise the group of flowerless and seedless plants that reproduce by means of asexual spores showing no differentiation into stem, root and leaf, and are deprived of chlorophyll.

genotype—the particular set of genes found within an organism.

genus—a category in biological classification comprising one or more phylogenetically related species.

germ theory—a theory in medicine that infectious diseases result from the action of microorganisms.

Golgi apparatus—a membrane-bound structure found within the cytoplasm of eukaryotic cells which functions in protein synthesis.

habitat—the type of environment in which an organism or group normally lives or occurs.

hantavirus pulmonary syndrome—a viral disease transmitted by exposure to rodent excrement via aerosol distribution, especially in moist areas, and rodent saliva from bites. This disease results in flu-like symptoms during the incubation period to severe respiratory problems, internal bleeding, and if the disease is not caught in time—death. The virus is divided into two groups: one found in Asia and Europe, the other in the United States. A recent outbreak in Chile in 1997 killed twenty-five people. From the initial outbreak in 1993 to 1997 a total of 172 cases werre reported in the United States with only forty-five percent resulting in death. Keeping an extremely clean house, especially the kitchen area, is key to the prevention of this virus. Other prevention measures include rodent proofing the area by keeping food in containers, discarding uneaten pet food, setting mouse traps, and keeping garbage properly stored. To date, there is no specific treatment for hantavirus, and the earlier the symptoms are reported, the better chance the patient has for recovery.

helix—something spiral in form.

helminth—a multicellular worm, that can be either free-living or parasitic; such as roundworms, tapeworms, or flukes.

hemagglutinin (H) protein—one of the two main proteins found on the surface of the virus which causes the flu; it is necessary for attaching the virus to the host cell.

hemorrhagic—showing evidence of bleeding; certain infections (hemorrhagic fevers) result in the loss of blood and body fluids.

hepatitis A, B, C—three types of this viral disease are transmitted in different ways: A, through ingestion of contaminated food or water; B, sexually transmitted, and the use of unsterilized needles; and C, transfusion of tainted blood or transplant of infected tissue, affecting the liver. Symptoms are nausea, fever, weakness, loss of appetite, and jaundice. Large nationwide outbreaks have occurred every decade, the last in 1989. Vaccinations have been developed for hepatitis A and B, but not for C. Screening of blood and organ tissue donors is the best means of prevention of hepatitis C.

hermaphroditic—having both male and female reproductive organs.

herpes—a recurrent viral infection caused by herpesvirus hominis (HVH); consists of the following five viruses: herpes simplex virus types 1 and 2, human cytomegalovirus, Varicella-Zoster virus, and Epstein-Barr virus.

heterotrophs—microorganisms which require carbon dioxide and other organic compounds for their nutrition and energy needs.

HIV (Human Immunodeficiency Virus)—a type of retrovirus that is responsible for the fatal illness Acquired Immunodeficiency Syndrome (AIDS).

hookworm—an intestinal parasitic infection caused by larval hookworms that penetrate the host's skin; heavy infection with hookworm can create serious health problems for newborns, children, and persons who are undernourished; hookworm infections occur mostly in tropical and subtropical climates and are estimated to infect about one billion people—about one fifth of the world's population.

host—an organism that provides food or shelter for another organism.

humoral immunity—the branch of the immune system in which antibodies are produced in response to a foreign antigen.

hypodermic—administered by injection beneath the skin.

icosahedral—having twenty equal sides or faces.

inflammatory disease—disease with inflamed tissue, characterized by pain, swelling, redness, and heat.

influenza—this virus, more commonly the "flu," is transmitted from the sneeze or cough of an infected person, person-to-person contact, or contact with objects that an infected person has contaminated with nose and throat secretions. The symptoms are fever, headache, chills, fatigue, muscle aches and pains, runny nose, sore throat, and hacking coughs. Flu shots have been developed, but some people are adversely affected by it. Good hygiene and sanitary measures are an effective means of prevention. People who do suffer from the virus are advised to drink plenty of fluids and rest. Influenza epidemics, such as the Spanish influenza of 1918, can result in massive devastation.

inhibitor—a molecule which represses or prevents another molecule from engaging in a reaction.

intravenous—occurring within or entering by way of a vein.

Junin virus—this South American arenavirus, known as the Argentine hemorrhagic fever, inflicts several hundred people annually; a rodent-borne virus whose origin or cause of spread remain unknown.

Kyasanur Forest disease—a tick-borne encephalitis complex; a flavivirus transmitted between infected ticks and monkeys which can cause severe hemorrhagic fever.

Kuru—a slow-virus disease rarely seen today because of the discontinuance of cannibalism and ritualistic butchering; caused neurodegenerative changes; symptoms included gait disturbance, incoordination, and swallowing difficulty.

Leishmanisis—a parasitic infection transmitted through the bite of a female sandfly; improving sanitary conditions and spraying insecticides are a means of prevention.

Leptospirosis—a bacterial infection that is transmitted through direct contact with water, food, or soil containing urine from an infected animal.

limbic system—a system of functionally related neural structures in the brain that are concerned with emotion and motivation.

lipid—a fat or fat-like substance which is insoluble in water but soluble in organic solvents, and is an essential structural component of living cells (along with proteins and carbohydrates).

Lyme disease—transmitted from the bite of a deer tick, this bacterium subjects the victim to a circular rash with a clear center area. Other early warning signs are flu-like symptoms: fever, headache, fatigue, and muscle or joint pains. Within weeks to months of the onset of the rash, more serious symptoms occur: heart abnormalities, meningitis, encephalitis, and facial palsy. The first reported case of Lyme disease in 1982 was in the Long Island area, and subsequent cases were reported in Hudson Valley, Ulster and Duchess Counties, and in Canada. Wearing protective clothing, long sleeves and putting pants inside the socks when outdoors, using insect and tick repellent containing the chemical DEET, are ways to prevent Lyme disease.

lymphocyte—a white blood cell present in the blood, lymph, and lymphoid tissue; the two major types are T cells and B cells.

lysosomes—structures found within the cytoplasm of certain eukaryotic cells which contain digestive enzymes; responsible for ridding the cell of debris.

macrophage—a large, immune system cell that devours foreign antigens and stimulates the action of other immune system cells.

macroscopic—large enough to be visible to the naked eye.

magnetic resonance imaging (MRI)—a special imaging technique used to image internal structures of the body, particularly the soft tissues; creating an image superior to a normal x-ray.

malaria—a tropical parasitic disease that kills more people than any other communicable disease except tuberculosis. It is reemerging in areas that had controlled or eradicated the disease, namely the Central Asian Republics of Tajikistan and Azerbaijan, and in Korea. The emergence of multi-drug-resistant parasitic strains has also caused the resurgence of this disease. Malaria is transmitted through the bite of an *Anopheles* mosquito, and if promptly diagnosed and adequately treated, is curable. Symptoms include high fever, severe chills, enlarged spleen, repeated vomiting, anemia and jaundice. Malaria endemic countries are some of the world's poorest. Antimalarial drugs work, but most of these countries can not afford them. Controlling the mosquito would be the best line of defense, through personal means—insect repellent, household areas, and community population methods—insecticides and environmental control.

measles—a viral infection that is spread through contact with the saliva from an infected person. Sharing utensils is the most common way to catch this contagious disease. Symptoms include rash, cough, and fever; childhood immunization is the best form of prevention.

metabolic reaction—chemical changes in living cells by which energy is provided for vital processes.

meningitis—an infection of the fluid of a person's spinal cord and the fluid that surrounds the brain. People sometimes refer to it as spinal meningitis. Meningitis is usually caused by a viral or bacterial infection. Knowing whether meningitis is caused by a virus or bacterium is important because the severity of illness and the treatment differ.

miasmatist—one who has made a special study of infectious particles or germs floating in the air.

microbe—a microscopic organism, such as a bacterium, a virus or a protozoan.

mitochondria—small intracellular organelles, found in eukaryotic cells, responsible for energy production and cellular respiration.

multicelluar—consisting of, or having, more than one cell or many cells.

multiple fission—splitting multiple times.

mumps—a virus that lives in the mouth, nose, and throat that can be transmitted when an infected person cough, sneezes, talks, or touches someone else. Symptoms include fever, headache, and swollen glands under the jaw; childhood immunization is the best form of prevention.

mutagen—an agent that can cause an increase in the rate of mutation; includes x-rays, ultraviolet irradiation, and various chemicals.

natural selection—a natural process that directs the evolution of organisms best adapted to the environment.

nematode—an unsegmented worm with an elongated, round body pointed at both ends; mostly free-living but some are parasitic.

nucleotide—the basic structural unit of nucleic acids (DNA or RNA).

nucleus—the membrane-bound structure found in eukaryotic cells which contains DNA and RNA and is responsible for growth and reproduction.

organelles—subcellular, membrane-bound structures, found within eukaryotic cells which perform discrete functions necessary for the life of the cell.

organism—any individual living thing, whether animal, plant, or microorganism.

outbreak—the occurrence of a large number of cases of a disease in a short period of time.

pandemic—an epidemic that affects multiple geographic areas at the same time.

paramecia—ciliate protozoa of the genus Paramecium that have an elongated body, rounded at the anterior end and a funnel-shaped mouth at the extremity.

parasite—an animal or plant that lives in or on another, and from which it obtains nourishment.

pasteurization—partial sterilization of food at a temperature that destroys harmful microorganisms without major changes in the chemistry of the food.

pathogen—any disease-producing agent, such as a virus, bacterium, or any other microorganism.

pertussis—a disease of the respiratory mucous membrane; also known as whooping cough.

phagocytosis—the intake of material into a cell by the formation of a membrane bound sac.

photosynthesis—the process by which green plants, algae, and some bacteria absorb light energy and use it to synthesize organic compounds (initially carbohydrates).

phototrophs—microorganisms (bacteria) capable of using light energy for metabolism.

phylum—the second-highest taxonomic classification for the kingdom Animalia (animals), between kingdom level and class level.

phylogeny—the evolutionary history of a particular taxonomic group.

pili—thread-like structures present on some bacteria; pili are shorter than flagella, and are used to adhere bacteria to one another during mating and to adhere to animal cells.

placebo—a medicinal preparation having no specific pharmacological activity against a patient's illness or complaint; given solely for the psychophysiological effects of the treatment.

plankton—small (often microscopic) plants and animals floating, drifting or weakly swimming in bodies of fresh or salt water.

plasmid—a small, independently replicating circle of DNA, found in bacteria, that can be transferred from one organism to another during certain types of mating.

polio—the virus gains entry to the body by fecal oral contact, or person to person contact. The disease causes paralysis, which is irreversible, and in more severe cases this paralysis can lead to death by asphyxiation. The symptoms are generally mild: low-grade fever, malaise, vomiting, stiff neck and back, and pain in the limbs. Large polio epidemics caused panic in the 1940s and 1950s in industrialized countries such as the United States and Western Europe. In 1954, Dr. Jonas Salk developed a vaccine that greatly decreased the occurrence of the disease, and in 1963 an oral vaccine was developed that led to its eradication in the United States. The virus is still prevalent in South Asia and Sub-Saharan Africa. Immunization is the best forms of prevention and treatment of polio.

polygyny—the mating of a single male with several females.

polymerase chain reaction (PCR)—the first practical system for in vitro amplification of DNA, and as such one of the most important recent developments in molecular biology.

polymorphonuclear leukocytes—also called neutrophils; white blood cells which respond quickly, phagocytose and destroy foreign antigens, such as pathogenic microorganisms.

polypeptides—two or more amino acids bound together which upon a chemical reaction with water yields multiple amino acids.

polysaccharide—any of a class of carbohydrates whose molecules contain chains of monosaccharide (simple sugar) molecules.

prions—an infective group of complex organic compounds (proteins) suggested as the causative agents of several infectious diseases.

proglottids—a segment of a tapeworm containing both male and female reproductive organs; capable of a brief independent existence.

prokaryotes—organisms, namely bacteria and blue green algae, characterized by the lack of a distinct nucleus.

prophylactic— preventive measure or medication.

protease— any enzyme that catalyses the splitting of interior peptide bonds in a protein.

protein— any of a group of complex organic compounds that contain carbon, hydrogen, oxygen, nitrogen, and usually sulfur, the characteristic element being nitrogen; widely distributed in plants and animals.

protist— unicellular, colonial, or multicellular organisms including protozoa and most algae.

protozoa— simple unicellular animals comprising some 50,000 organisms.

pseudopods— temporary blunt ended projections of the cytoplasm of a cell that is used for locomotion or food collecting (in amoeba).

Q fever— a bacterial infection that is contracted by contact with materials contaminated with animal feces, blood, inhaling contaminated dust or droplets, or ingesting contaminated food or liquids receptor—a molecule on the surface of a cell that serves as a recognition or binding site.

recombination— formation by the process of crossing over an independent assortment of new genes in the offspring that did not occur in the parents.

reservoir host— a host that carries a pathogen without injury to itself and serves as a source of infection for other host organisms.

ribosomes— a structure found within the cytoplasm of cells, made up of protein and RNA, that serves as the site of protein synthesis.

Rift Valley fever— an acute, fever-causing viral disease that affects domestic animals (such as cattle, buffalo, sheep, goats, and camels) and humans; RVF is most commonly associated with mosquito-borne epidemics during years of heavy rainfall.

Rinderpest virus (RPV)— a highly infectious viral disease that can destroy entire populations of cattle and buffalo; the only way to prevent this disease is to vaccinate all animals and livestock.

river blindness— a parasitic worm disease that is spread by the bite of a blackfly; the best way to prevent the disease is insect repellent.

RNA (ribonucleic acid)— a nucleic acid that governs protein synthesis in a cell.

serum— the clear, thin fluid portion of the blood which remains after coagulation; antibodies and other proteins are found in the serum.

spore formation— formation of unicellular, often environmentally resistant, dormant or reproductive bodies produced by some microbes.

sexually transmitted disease (STD)— a communicable disease transmitted by sexual intercourse or genital contact.

symbiotic— the relationship between two interacting organisms or populations.

T cells— thymus-derived white blood cells (lymphocytes) that participate in a variety of cell-mediated immune responses.

T -cytotoxic cells— a subset of T lymphocytes which are able to directly kill foreign cells, especially virally infected host cells.

T -helper cells— a subset of T lymphocytes which normally orchestrate the immune response by signaling other cells in the immune system to perform their special functions.

tegument— the covering of a living body, or of some part or organ of such a body.

thymus— the lymphoid organ in which T lymphocytes are educated, mature and multiply.

Toxigenic E. coli— a bacterial infection transmitted through the ingestion of undercooked ground beef, unpasturized milk, or water that has been contaminated by sewage. Symptoms include bloody diarrhea, abdominal cramps, nausea, fever, and occasionally kidney failure. A recent outbreak was in June of 1998 in the United States affecting approximately 4,500 people. Other outbreaks have been reported in Japan and England. The disease can be prevented through sanitary measures such as washing hands, cooking meat thoroughly, and avoiding unpasturized milk.

Trypanosomiasis (African sleeping sickness)—a parasitic infection that is spread through the bite of the tsetse fly. The bite itself is very painful, and in the early stages a red sore appears at the site of the bite. In the weeks to follow, the patient experiences fever, rash, extreme fatigue, swelling around the eyes and hands, muscle and joint aches, severe headaches, swollen lymph nodes, and weight loss. As the disease progresses, the central nervous system is invaded, and if left untreated, death will occur. The reason it is known as the sleeping sickness is because the person infected by the parasite will sleep during the day, all day, and experience insomnia at night. Immediate treatment and some hospitalization is the way to treat the disease. Found only in Africa, there was a resurgence of the disease in the southern Sudan in 1997. The best defense is to guard against the bite of tsetse fly—wear protective clothing made of thick fabric and olive or khaki in color because the tsetse fly is attracted to bright colors, avoid bushy areas, sleep with bednetting, and the use insect repellent.

unicellular—having only one cell.

vaccine—a substance that contains antigenic components, either weakened, dead or synthetic, from an infectious organism which is used to produce active immunity against that organism.

vector—an organism that transmits a pathogen.

vertebrate—one of the grand divisions of the animal kingdom, comprising all animals that have a backbone composed of bony or cartilaginous vertebrae.

virion—a single virus particle, complete with coat .

virulence—the degree or ability of a pathogenic organism to cause disease.

virus—ultramicroscopic infectious agent that replicates itself only within cells of living hosts.

white blood cell—white corpuscles in the blood; they are spherical, colorless masses involved with host defenses; blood cells that engulf and digest bacteria and fungi; an important part of the body's defense system.

World Health Organization (WHO)—an agency of the United Nations founded in 1948 to promote technical cooperation for health among nations, carry out programs to control and eradicate disease, and strive to improve the quality of human life.